SpringerBriefs in Computer Science

For further volumes:
http://www.springer.com/series/10028

Katia Potiron · Amal El Fallah Seghrouchni
Patrick Taillibert

From Fault Classification to Fault Tolerance for Multi-Agent Systems

 Springer

Katia Potiron
Thales Airborne Systems
Louveciennes, Yvelines
France

Patrick Taillibert
Thales Airborne Systems
Louveciennes, Yvelines
France

Amal El Fallah Seghrouchni
LIP6, University Pierre and Marie Curie
Paris
France

ISSN 2191-5768   ISSN 2191-5776 (electronic)
ISBN 978-1-4471-5045-9   ISBN 978-1-4471-5046-6 (eBook)
DOI 10.1007/978-1-4471-5046-6
Springer London Heidelberg New York Dordrecht

Library of Congress Control Number: 2013932457

Printed on acid-free paper

Springer is part of Springer Science+Business Media (www.springer.com)

Preface

This book presents a summary of the research achieved by Katia Potiron during her Ph.D. Thesis defended on 14 April 2010 at the University Paris 6. This thesis was co-supervised by Patrick Taillibert, expert engineer at Thales Airborne Systems and Amal El Fallah Seghrouchni, Professor at the University Paris 6 (University Pierre and Marie Curie, France).

The intuition of this research is that, thanks to the Multi-Agent Systems (MAS) approach, fault tolerance property might be achieved naturally for complex software and should be complementary with classical existing approaches. This book, as a one-year "night work" by Katia Potiron after her thesis defense, tries to summarize the thesis contributions and to provide a comprehensive view of this research. The book contains additional explanations to make accessible to those not familiar with the subject the ideas sustaining the thesis propositions.

From a research point of view, this work is a cross-disciplinary attempt between the well-established field of fault tolerance and the emerging field of MAS issued from distributed artificial intelligence. Indeed, the MAS paradigm plays today an important role in complex software development. The associated technology offers a large panel of original concepts, architectures, interaction protocols, and methodologies for the analysis and the specification of complex systems built as MAS. One of the driving motivations for this work is the observation that MAS, as a technology, still lacks mechanisms to guarantee robustness and fault tolerance properties. These properties are crucial from a software perspective, especially when MAS are built for critical or military applications where dependability is vital. The expected properties vary according to the effects of the abnormal behavior of the software on the system safety, what is represented, for example, by design assurance level for software in civil airborne systems but this aspect is not studied here since it is related to a third domain that would be safety assessment. Hence, this book tries to emphasize the characterization of MAS with regard to existing studies in fault tolerance domain.

For classical systems, a fault classification exists and allows defining faults. So that, when dependability is at stake, such a fault classification may be used, from the beginning of the system design, to define fault classes and specify which types of faults are expected for the system and the software. Thus, one may tend to use such a fault classification for MAS, but the fact that agents are autonomous and proactive may come into consideration on the faults potentially occurring in the system. As a matter of fact, this kind of behavior is not taken into account in the present fault classification. Moreover, autonomous and proactive agents are primarily "intelligent agents". Does this "intelligence" have a role to play with regard to fault tolerance? Is it possible to take advantage of an agent property to obtain a more effective fault handler? Or are the agent properties an impediment to fault tolerance?

In addition, the field of fault tolerance provides numerous methods adapted to handle different kinds of faults. Some handling methods had been studied in the Multi-Agent System domain, adapting to their specificities and capabilities but at the same time increasing the large amount of fault tolerance methods to consider. Therefore, unless one is an expert in fault tolerance, it is difficult to choose, evaluate, or compare fault tolerance methods. This prevents many applications from using these methods and, consequently, to be tolerant to common faults. That is the reason why this book also tries to answer the important question of how to derive some guidelines and fault handlers based on the fault classification and the MAS studies (for instance from the properties specification phase).

Finally, the authors would like to thank Gilles Klein for his insightful review of this book, Costin Caval for his review and future work on handling unforeseen faults, and Nicolas Viollette for his long-term support to Katia Potiron. We also address a special acknowledgment to Professor Karin K. Breitman who made possible the publication of this thesis.

December 2012   Katia Potiron
                                                          Amal El Fallah Seghrouchni
                                                                  Patrick Taillibert

Contents

Chapter 1
Introduction

Abstract This chapter introduces the background and goal of this book providing an outline of the following sections.

Keywords Multi-agent system · Dependability · Fault classification · Autonomy · Multi-agent system design

In artificial intelligence research, Multi-Agents Systems (MAS) technology has been praised as a new paradigm for conceptualizing, designing, and implementing software systems.

When studying dependability, which is essential for the use of MAS for industrial and military software, an important point is to know which faults the system will experience.

The importance of this knowledge comes from the necessity to decide and define precisely, during the system definition phase, which kinds of faults should be taken into consideration. This fact is underlined by Gartner in [1]: "When designing fault tolerance, a first prerequisite is to specify the faults that should be tolerated... what is done more concisely by specifying a fault class".

The specifications definition of the system will not only consider the faults with regard to define detection means and handlers needed for fault tolerance (that represents all means to avoid service failures when faults are involved) but also, at many stages of the system life cycle:

- during system design and construction:

  - for fault prevention;
  - for fault removal;
  - for fault forecasting;

- during system use:

  - for fault evaluation.,

  e.g. faults are considered during all the system lifetime for all the means of dependability [2].

K. Potiron et al., *From Fault Classification to Fault Tolerance for Multi-Agent Systems*,
SpringerBriefs in Computer Science, DOI: 10.1007/978-1-4471-5046-6_1,
© The Author(s) 2013

Considering the central influence of faults on the design of a dependable system, it is necessary to adopt a common basis to exchange knowledge, to specify systems or to compare methods. For these reasons some work has been done for a long time on the taxonomy issue for fault classification in software systems but it came to our attention that no work has been specifically done for MAS.

The fact that fault classification has not been studied in depth for MAS is particularly striking when considering autonomy and pro-activeness. These two major characteristics of agents clearly distinguish MAS from other software systems. This book does not address the subject of defining autonomy neither pro-activeness. We consider that *pro-activeness represents the fact that agents have internal goals and own activities*. For autonomy definition, we will use a point that most of its definitions have in common: *autonomy allows agents to take their decisions on their own, e.g. "without direct human or other intervention"*.

From a software engineering point of view, autonomy can be perceived as the fact that the agent designer does not know exactly the specifications that are the internal laws and/or the internal state of the other agents. And moreover, during their execution agents will only have a limited perception of their environment not allowing them to predict other agents behavior. From this point of view, MAS are a special case of distributed systems, characterized by agents autonomy and pro-activeness.

Faults are a concern for MAS designers, especially because agents are interacting with unpredictable agents. To deal with faults, the designer must have some precise information on the faults MAS are subject to.

After defining the faults MAS are subject to, it came to our attention that some of these faults lack a generic handler for agents. The study of this fact results on a new fault handler adapted to autonomy, pro-activeness and "intelligence" of agents. We provide a definition of this handler and a detailed feedback of the use of this fault handler to design robust agents.

Our final goal is to find a way to build MAS where fault tolerance is naturally a property of agents and platform (i.e. achievable directly at the creation of the program and without a specific effort from the designer). To obtain such a good property, fault tolerance must be "made for MAS" to take into account specificities of MAS, particularly the fact that agents are high level entities. Toward this objective, our work paves the way for designers to specify fault tolerance for platforms and agents into MAS.

The book will be organized as follows. Chapter 2 summarizes definitions and properties of multi-agent systems. Chapter 3 presents a fault classification and Chap. 4 our analysis and adaptations of this fault classification for MAS. Chapter 5 presents a study and proposition for a fault handler adapted to multi-agents systems specific faults. Chapter 6 presents the issue of building fault tolerant MAS describing analysis of platform and agents and then analysis of faults handlers. The last chapter concludes this work and presents some perspectives.

References

1. Gartner, F.C.: Fundamentals of fault-tolerant distributed computing in asynchronous environments. ACM Comput. Surv. **31**(1), 1–26 (1999)
2. Avizienis, A., Laprie, J.-C., Randell, B., Landwehr, C.: Basic concepts and taxonomy of dependable and secure computing. IEEE Trans. Dependable Secur. Comput. **1**(1), 11–33 (2004)

Chapter 2
Multi-Agent System Properties

Abstract This chapter succinctly presents Multi-Agent Systems (MAS) as software architecture composed of entities: agents, presenting specific properties: autonomy and pro-activeness; and a platform providing agents for communication means. These properties are then analyzed with regard to their impact on possible faults and impacts on system dependability.

Keywords Multi-agent system · Autonomy · Pro-activeness · Adaptability · Dependability · Non deterministic behavior

2.1 Multi-Agent System and Agents

A multi-agent system is a set of software agents that interact to solve problems that are beyond the individual capacities or knowledge of each individual agent. We call a "platform" whatever allows those agents to interact not taking into consideration the shape that such a platform can take (centralized or not, embedded into the agents or not, ...). This platform usually provides agents with a set of services depending on the system needs and is considered as a tool for the agents, it does not exhibit an autonomous or pro-active behavior.

Agents, according to MAS community (see for example [1]) have the following properties:

- *Autonomy*. An agent possesses individual goals, resources and competences; as such it operates without direct human or other intervention, and has some degree of control over its actions and its internal state. One of the foremost consequences of agent autonomy is agent adaptability as an agent has the control over its own state and so can regulate its own functioning without outside assistance or supervision.
- *Sociability*. An agent can interact with other agents, and possibly humans, via some kind of agent communication language. Through this means, an agent is able to provide and ask for services.

K. Potiron et al., *From Fault Classification to Fault Tolerance for Multi-Agent Systems*, SpringerBriefs in Computer Science, DOI: 10.1007/978-1-4471-5046-6_2,

- *Reactivity*. An agent perceives and acts, to some degree, on its close environment; it can respond in a timely fashion to changes that occur around it.
- *Pro-activeness*. Although some agents, called reactive agents, will simply act in response to stimulations from their environment, an agent may be able to exhibit goal-directed behavior by taking the initiative.

Even if we will not address the subject of defining the autonomy, we will study the consequences of a point that most definitions have in common. The point is that autonomy allows agents to take their decisions on their own, e.g. "without direct human or other external intervention".

For a commonly accepted definition of agent autonomy, we would like to mention Henry Hexmoor's [2]:

"An agent is autonomous with respect to another agent, if it is beyond the influences of control and power of that agent."

As well as the definitions given in [3, 4].

These definitions of agent autonomy implies that agents are not entitled to dictate others decisions or actions. Those are huge hypothesis for the agents that can lead to an unproductive system if no agent would like to work with others and must be balanced with the collaborative aspects of agents. This can be seen as: "I will work with you, not because you have the power the force me to but because of mutual interest and if I have a more profitable offer I will stop working with you".

Moreover agents do not know the decision criteria of the other agents neither during execution (agents are distributed and partitioned entities and so have no access to internal variables of others) nor than during conception (interactions are defined but the intended behavior of the other agents are not known). Those are every day condition when building a software system; more and more entities of the system are not built by the same team or company (and in some cases not in the same country).

As a consequence, agents gain some independence with regard to the other agents; they can go on and thus survive even if no other agent is available.

This aspect of autonomy, decision criteria are internal and private to the agent, provide agent with more independence with regard to the other agents (they can go on and survive even if other agents are unavailable. This contributes to system robustness through the modularity and fault isolation aspects.

Nonetheless, autonomy makes the behavior of agents not completely foreseeable and therefore non-deterministic. This unpredictability is a property of interacting entities; therefore, agents taking "their decisions by themselves" (i.e. by taking non-supervised decisions on partial knowledge[1]) can, whether voluntarily or not, be responsible for faults that other agents will experience.

---

[1] Note that the "partial knowledge" consideration is directly a part of *Reactivity* definition given by [1].

This faulty aspect of agent autonomy had also been underlined by Hägg in [5]:

"When discussing fault handling in multi-agent systems (MAS) there are often two opposite positions. The first says that a MAS is inherently fault tolerant by its modular nature; faults can be isolated and generally do not spread. The other says that a MAS is inherently insecure, as control is distributed. It is non-deterministic, and a specific behavior is hard to guarantee, especially in fault situations."

This distributed way of controlling the application induces the fact that agents undergo the autonomous behavior of other agents in two different ways:

- One agent may choose (using its internal methods and private variables) not to respond to a request from another agent for any reason it considers acceptable (suspicion toward the other agent, more important things to do, schedule overlap, being overloaded...);
- Malicious agents can choose to harm other agents. From a software engineering point of view, autonomy can be perceived as the fact that the agent designer does not know exactly the specifications, internal laws or internal state of other agents during their execution. Even if MAS are distributed systems, autonomy is the breaking point. Moreover, in some MAS (called open MAS), even the developers (and owners) of the other agents may not be known.

Note that form the "game" point of view such a malicious and harming agent may not be a *faulty* agent but a *winning* agent, if harming other agents allows itself to obtain a benefit. In a competition environment, autonomy and pro-activeness of agents may be expressed in various ways.

2.2 Agents and Faults

As agents know that other agents are autonomous and then unpredictable, they must take their decisions on their own and be able to adapt themselves to changing situations. Doing so, they gain independence or a degree of freedom with regard to their relations to other agents. They can go on and thus survive even if other agents are not available. This aspect of autonomy facilitates two aspects of fault tolerance:

- Error confinement, since the isolation of one or more suspected faulty agent(s) may not result in totally incapacitating the other agents and,
- System readjustment, since agents considering that others are autonomous must have adaptability capacities that result in the fact that a loss of some agents will not imply a complete loss of service for the MAS.

Those two aspects make community of agents (and so Multi-Agent Systems) more robust.

But at the same time, it is also said that autonomy makes fault tolerance a more difficult task—see [5, 6]—and one can wonder why such a statement was expressed.

This statement is a direct consequence of the decision-making process resulting from agent autonomy: the behavior of an agent is not completely foreseeable for the agents interacting with it, which requires for the agent to be prepared to adapt its behavior to unforeseen situations. Considering that this adaptation is needed for every decision and action of other agents results in the creation of a lot of unpredictable situations that should be handled. This aspect does not facilitate agents design.

It is then clear that agents cannot be designed like classical systems and fault tolerance must have a particular place in their design. As a matter of fact, since autonomy and pro-activeness are a natural feature of agents, faults induced by their behavior cannot be "removed" like it may be the case for development faults in monolithic deterministic software. This became even more complicated when emerging behaviors are taken into consideration.

Autonomy, *limited environment sensing* and *adaptability* are interesting properties since they correspond to the way software development is evolving and make multi-agent systems relevant even in industrial or military domains like airborne systems.

Software systems being more and more complex imply that:

- Safety studies cannot predict every possible harmful behavior of the system.
- Software modules of a system cannot be aware of all actions of other modules.

Then reconfiguration is one of the few adapted solution for the system to be able to have a better continuity of service.

This book tries to answer the questions rising from these considerations and to provide multi-agents system with new arguments for their use in industrial or military domains:

- What is to be done to specify and implement an agent interacting with such unpredictable entities?
- What is to be done to specify and implement the whole MAS?

2.3 Agents and Fault Tolerance

Some work on MAS has been done with regard to fault tolerance domain that provides agents and MAS designers with responses to these questions.

Two main line of research in the MAS domain are addressing fault tolerance (we exclude here the ad hoc adaptation of interaction protocols and non reusable methods), they are presented next.

One line of research for agents fault tolerance is to include the fault tolerance directly into the languages representing the behavior of the agent. The creators of Cool [7] or AgenTalk [8] in their chapters introducing their respective languages also present some prospects on mechanisms to handle faults into the conversations.

They propose to use a meta script to control the execution of the behavioral scripts, the meta scripts being started with messages, representing the exceptions, sent from the behavioral script to the meta script. The authors do not provide any general meta script of control and assume that they will be built in a suitable way for each exception. They conclude saying that the meta scripts may be grouped together using exception classes.

More recently, Dragoni and Gaspari [9] proposed a fault tolerant agent communication language (FT-ACL) that deals with crash failure of agent, providing fault tolerant communication primitives and support for an anonymous interaction protocol. The knowledge level of the FT-ACL is based on giving for each speech act a success continuation and a failure continuation.

These methods are focused on conversational agents and on the handling of faults perceived into the communications but the realization of the suitable handler is the responsibility and the burden of the agent designer/developer. The more serious issue of such a handler creation is the risk for the developer to introduce new faults into its handler.

Another line of research in MAS, on exception handling for MAS uses sentinels [5, 10–12] to diagnose (from the analysis of the received and sent messages) and handle exceptions for agents. This approach is different from the approaches presented previously since the diagnosis and decision about the handler adapted to the fault is outside the agent. The choice of the handler can be done inside a knowledge-base containing specific handlers [10]. Besides the creation of the external entity, it requires some special design of the agents to allow an external entity to change its internal state.

And there are also other lines of research in MAS fault tolerance like for example:

- The replication (or cloning) of agents is used in [13, 14] to deal with physical faults.
- "Prevention of harmful behaviors" [15] which deals with the emergence of harmful behaviors of agents.

Several lines of research in MAS never refer to fault tolerance but provide valuable strategies for it, like:

- Computational trust and reputation [16] to deal with malicious agents,
- Agent planning [17, 18] to deal with unexpected situations.

In all these research works, it appears that the kind of faults taken into consideration is never clear and hence the efficiency of the proposed fault tolerance solutions with regard to fault handling is difficult to estimate. A way to address these questions is to study faults in order to classify them in a limited number of fault classes and this is indeed what will be presented in the next chapter.

References

1. Wooldridge, M., Jennings, N.R.: Intelligent agents: Theories, Architectures and Languages, January 1995. Lecture Notes in Artificial Intelligence, vol. 890, ISBN 3-540-58855-8
2. Hexmoor, H.: Stages of autonomy determination. Trans. Sys. Man Cyber. Part C **31**(4), 509–517 (2001)
3. d'Inverno, M., Luck, M.: Understanding autonomous interaction. In: Wahlster, W. (ed.) Proceedings of the 12th European Conference on Artificial Intelligence, pp. 529–533. John Wiley and Sons (1996)
4. Castelfranchi, C., Falcone, R.: From automaticity to autonomy: the frontier of artificial agents. In: Hexmoore, H., Castelfranchi, C., Falcone, R. (eds.) Agent Autonomy, Multiagent Systems, Artificial Societies, and Simulated Organizations, vol. 7, pp. 103–136. K. Academic (2003)
5. Hägg, S.: A sentinel approach to fault handling in multi-agent systems. Paper presented at Second Australian Workshop on Distributed AI in conjunction with the Fourth Pacific Rim International Conference on Artificial Intelligence, pp. 181–195. Springer-Verlag, London (1996)
6. Zhang, Y., Manisterski, E., Kraus, S., Subrahmanian, V.S., Peleg, D.: Computing the fault tolerance of multi-agent deployment. Artif. Intell. **173**(3–4), 437–465 (2009)
7. Barbuceanu, M., Fox, M.S.: Cool: a language for describing coordination in multiagent systems. In: Lesser, V., Gasser, L. (eds.), Proceedings of the First International Conference oil Multi-Agent Systems, June 12–14, 1995, San Francisco, California, USA, pp. 17–24, The MIT Press (1995)
8. Koren, I., Koren, Z., Su, S.Y.: Analysis of a class of recovery procedures. IEEE Trans. Comput. **35**(8), 703–712 (1986)
9. Dragoni, N., Gaspari, M.: Crash failure detection in asynchronous agent communication languages. Auton. Agent. Multi-Agent Syst. **13**(3), 355–390 (2006)
10. Klein, M., Dellarocas, C.: Exception handling in agent systems. In: Etzioni, O., Müller, J.P., Bradshaw, J.M. (eds.) Proceedings of the Third International Conference on Autonomous Agents (Agents'99), ACM Press, Seattle, pp. 62–68 (1999)
11. Platon, E., Sabouret, N., Honiden, S.: A definition of exceptions in agent oriented computing. In: O'Hare, G., O'Grady, M., Dikenelli, O. Ricci, A. (eds.) Proceedings of the 7th International Conference on Engineering Societies in the Agents World VII, pp. 161–174, Springer-Verlag, Berlin, Heidelberg (2007)
12. Shah, N., Chao, K.-M., Godwin, N., James, A.E.: Exception diagnosis in open multi-agent systems. In: IAT, pp. 483–486 (2005)
13. Fedoruk, A., Deters, R.: Improving fault-tolerance by replicating agents. In: Proceedings of the First International Joint Conference on Autonomous Agents and Multiagent Systems: part 2, ACM Press, Bologna, Italy, pp. 737–744. (2002)
14. Guessoum, Z., Faci, N., Briot, J.-P.: Adaptive replication of large-scale multiagent systems: towards a fault-tolerant multi-agent platform. In: Proceedings of the Fourth International Workshop on Software engineering for Lrge-Scale Multi-Agent Systems, ACM Press, St. Louis, Missouri, pp. 1–6. (2005)
15. Chopinaud, C., Fallah-Seghrouchni, A.E., Taillibert, P.: Prevention of harmful behaviors within cognitive and autonomous agents. Paper presented at European Conference on Artificial Intelligence, pp. 205–209 (2006)
16. Sabater, J., Sierra, C.: Review on computational trust and reputation models Artif. Intell. Rev. **24**(1), 33–60 (2005)
17. de Weerdt, M., ter Mors, A., Witteveen, C.: Multi-agent planning: an introduction to planning and coordination. In: Handouts of the European Agent Summer School, pp. 1–32, (2005)
18. Seghrouchni, A.E.F., Hashmi, M.A.: Multi-agent planning. In: Essaaidi M. et al. (eds.) Software Agents, Agent Systems and Their Applications. IOS Press (NATO Science for Peace and Security Series) (2012)

Chapter 3
Fault Classification

Abstract Chapter 2 outlined that agent can make faults and suffer from faults like any other system and that autonomy and pro-activeness properties of agents are a source for agent faults. Moreover some specific research exists on fault tolerant agents. What we were looking for then was a tool to get the measure of those faults and fault tolerance methods with regard to multi-agents systems needs. This chapter presents, first, a conventional fault classification. Then this conventional fault classification is used as the basis for the analysis, presented in a second part, of applicability of the conventional fault classification considering the faults observed by multi-agent systems designers.

Keywords Multi-agent system · Dependability · Fault classification · Autonomy · Behavioral faults

3.1 Conventional Fault Classification

Let us first recall several definitions about faults, errors and failures [1], that are illustrated on Fig. 3.1:

- Faults are generally defined as judged or hypothesized causes of an error.
- Errors are deviations of one or more of the external state of the system from correct service state, they have direct effects on the system.
- Failures are events that occur when the delivered service deviates from the correct one.

These are the definition of faults, errors and failures that will be considered for the rest of this book.

The study of fault classification presented in [1, 2] began in the early 1980s. To establish their classification, the authors studied a wide group of faults including short-circuits in integrated circuits, programmers' mistakes, electromagnetic

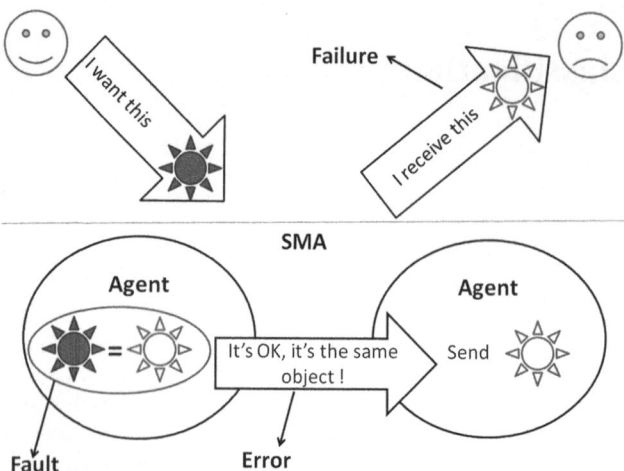

Fig. 3.1 Illustration of "fault/error/failure" propagation

interferences or inappropriate man/machine interactions. Other classifications can be found for example in [3] where authors classify faults by duration, nature and extent. Every fault classification we know of can be considered as being a sub-classification of the one defined by Avizienis et al. in [1, 2].

In [2], faults are classified according to seven attributes[1]:

- Phase of creation or occurrence,
- System boundaries,
- Dimension,
- Phenomenological cause,
- Objective,
- Capability,
- Persistence.

Each attribute has a set of exclusive values (for instance, values associated with the attribute *phase of creation or occurrence* are: *development* or *operational*), as shown on Table 3.1 which also provides the associated definitions.

A fault is then described as a complete assignment of a single value to each attribute, i.e. a 7-uple constituted of a value for every attribute entry in the table.

In this way, the seven attributes and their values produce 192 possible combinations.

---

[1] In [1], there were eight attributes as *capability* was separated into: (1) *intent* with *deliberate* and *non-deliberate* faults as values and (2) *capability* with *accidental* and *incompetence* faults as values. But the *intent* viewpoint appeared redundant and so authors chose to withdraw it.

Table 3.1 Attributes and their values used to describe fault classes [1]

Attributes	Values	Definitions
Phase of creation or occurrence	Operational	Occur during service delivery of the use phase
	Development	Occur during: (a) System development including generation of procedures to operate or to maintain the system (b) Maintenance during the use phase
System boundaries	Internal	Originate inside the system boundary
	External	Originate outside the system boundary and propagate errors into the system by interaction or interference
Dimension	Hardware	Originate in, or affect, hardware
	Software	Affect software, i.e., programs or data
Phenomenological cause	Natural	Caused without human participation
	Human-made	Result from human participation
Objective	Malicious	Introduced with the malicious objective of causing harm to the system
	Non-malicious	Introduced without a malicious objective
Capability	Accidental	Introduced inadvertently
	Deliberate	Result of a decision
	Incompetence	Result from a lack of professional competence by the authorized human(s), or from inadequacy of the development organization
Persistence	Permanent	Presence is assumed to be continuous in time
	Transient	Presence is bounded in time

The authors did not retain all of these combinations, because not all are relevant. Thus, a fault cannot be malicious and accidental at the same time. The remaining 25 possible fault classes, called *conventional faults* in this book, are illustrated in the fault classification tree presented on Fig. 3.2.

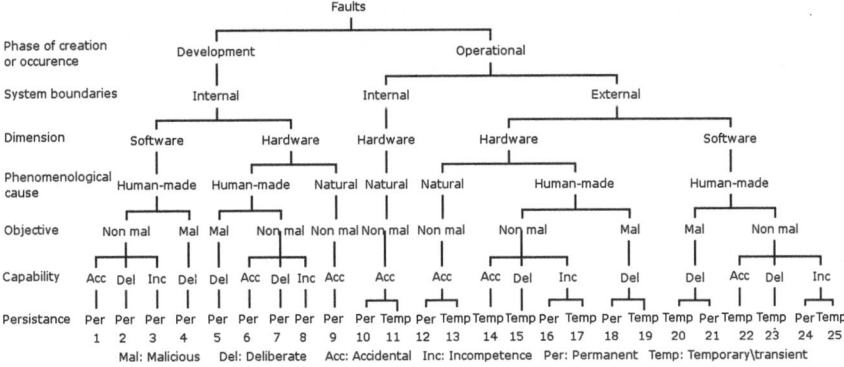

Fig. 3.2 Fault classes combinations

Table 3.2 Groups of faults and examples of terms definition

Development faults																			Interaction Faults					
									Physical faults															
1	2	3	4	5	6	7	8	9	10	11	12	13	14	15	16	17	18	19	20	21	22	23	24	25
Software flaw			Logic bomb		Errata hardware / Production defects				Physical Deteri-oration		Physical interference								Intrusion	Virus	Input mistakes			

It is then possible to characterize the large number of everyday or specialized terms denoting faults as the set of fault classes they belong to. Some illustrative examples of fault names and corresponding fault classes are given in the boxes at the bottom of Table 3.2.

The faults are also shown to belong to three major non-exclusive groups representing some practical points of view (first two lines of Table 3.2), described as follows:

- Development faults:

  - All fault classes with the value *development* for the *phase of creation or occurence* attribute.

- Physical faults:

  - All fault classes with the value *hardware* for the *dimension* attribute.

- Interaction faults:

  - All fault classes with the value *external* for the *system boundary* attribute.

The knowledge of every possible fault classes is a mean for system or agent designer to decide, during the system or agent specifications or during a safety study, which fault classes should be taken into consideration. For systems where not all fault classes are applicable (for example, physical deteriorations are not taken into consideration for software you have on your computer at home) this fault classification is a fast way to obtain an opinion of the customer on the faults that may need to be tolerated. For systems where safety is more important a complete safety assessment should be done but the fault classes and groups remain a good ontology for discussions.

We will next investigate how this fault classification complies when applied to MAS and how it faced some faults mentioned in MAS domain. As a matter of fact, the fault classification presented here is suitable for systems including software systems (and so for MAS that are software systems).

But, as mentioned in the introduction, MAS specificities lead us to refine the conventional fault classes in order to meet these specificities.

3.2 Existing Faults in MAS

3.2.1 Conventional Faults Relevance for MAS

MAS contains separate components interacting with each other to achieve a goal. This implies the following:

- Since they are made of software modules, MAS are vulnerable to all kinds of development faults;
- Since they are interacting programs, MAS are vulnerable to all kinds of physical faults;
- Since they are composed of interacting pieces of software (agents), MAS are vulnerable to all kinds of interaction faults.

Through their relationship with classical systems (and particularly with classical software systems), MAS inherit the conventional faults. The classification tree presented into Fig. 3.2 is therefore entirely relevant for MAS.

But, this classification being necessary does not make it sufficient for MAS. When MAS are concerned autonomy and proactivity, as mentioned in the introduction and previous section must be taken in consideration. So comes the next question we will address: Do studies of fault tolerance in MAS define faults that do not match the previously presented fault classes? Does the MAS fault-tolerance domain define specific faults?

3.2.2 Reported Handled Faults in MAS

As we have already explained in Sect. 2.3, MAS fault tolerance has been addressed using various methods:

- "Exception handling" [4, 5] provides agent architectures to deal with exceptional situations,
- "Agent replication" [6, 7] provides a fault tolerance method particularly suited to physical faults,
- "Sentinels" [4, 5, 8] provides agents with an external supervisor entity,
- "Prevention of harmful behaviors" [9] allows the definition of laws the agent will be constrained to respect,
- "Fault tolerant agents communication language" [10] deals with crash failure detection.

In addition, some other lines of research in MAS do not refer to fault tolerance but provide valuable strategies for it, as computational trust and reputation [11] to deal with malicious agents or planning [12] to deal with unexpected situations.

The MAS domain offers various prospects to fault tolerance but in all these researches, it is not clear what kinds of faults are taken into consideration and hence how efficient they are with regard to fault tolerance.

Moreover, even if faults are a concern for MAS designers, not many authors tried to provide the fault classes addressed by their fault tolerance methods. Whereas some examples of the faults defined in works for MAS fault tolerance exist and are discussed next.

In [13], Hägg considers four main sources of faults defined according to Burns and Wellings definitions:

- Inadequate specification of software
- Software design error
- Processor failure
- Communication error

In [6], the authors give some fault types handled by their replication method with some corresponding definitions:

- Program bugs:

  - Programming errors not detected by system testing;

- Unforeseen states:

  - Omission errors in programming, the programming does not handle a particular state;

- Processor faults:

  - System crash or a shortage of system resources;

- Communication faults:

  - Slow downs, failure or other problems with the communication links;

- Emerging unwanted behavior:

  - System behavior which is not predicted.

In [8], the authors define exception sources as belonging to one of these three levels:

- Environmental:

  - IO problem, network communication, CPU, or program errors;

- Knowledge:

  - Unable to understand the terms of the message;

- Social:

  - Agent assumptions about the coordination protocol are violated.

But these are *ad hoc* definitions, lacking general agreement about the concerned faults. It is neither obvious whether they address the same faults, nor which one may be best adapted to handle which faults.

Again, a way to determine the meaning of these terms is to consider the corresponding faults in existing fault classification systems.

But, when considering these faults, the outcome is not what can be expected: some faults do not match the classification, as shown in Table 3.3.

Emerging unwanted behavior and *social faults* do not correspond to any fault classes in the existing fault classification. They cannot be classified according to any value of the first attribute. This is due to the fact that they are linked to agent *autonomy*, not with *development* faults or with *service delivery*.

Consider the following example:

An agent A fails to provide a service to its user after not receiving the expected response from an agent B and this because B has autonomously evaluated that sending a response to A was not interesting.

With regard to the given definitions of faults (see Sect. 3.1): *faults are generally defined as judged or hypothesized causes of an error that then creates a failure.* The previous example can be analyzed as follows:

- The failure is that there was no received message in response to agent A request;
- The error is that agent A was not able to send its own response to the user;
- The fault, the root reason, is that agent B has not responded to agent A.

Agent B not sending a response will be a fault from agent A point of view and not so from agent B point of view because its autonomy allows it to take its decisions on its own and it is authorized to refuse a request even if from a social point of view agent B may have responded to agent A. Example, agent B may have some good reasons not to respond to agent A, it may have made him not able to respond to a more important request but agent A does not know this.

Table 3.3 Correlation map between fault in MAS and the fault classification

Names of the faults	Corresponding faults in Fig. 3.1
Program bugs [6] Inadequate specification of software [13] Software design error [13] Unforeseen states [6] Knowledge faults [8]	1–4
Processor faults [6] Processor failure [13]	5–9
Communication faults [6] Communication error [13]	10–25
Environment faults [8]	12–25
Emerging unwanted behavior [6] Social faults [8]	No match with conventional faults

Such a fault corresponds to a *software* and *non-human-made* fault:

- *Software*: because agent B is an autonomous piece of software and is functioning correctly,
- *Non-human-made*: because agent B is autonomous and so allowed to take this decision so agent B decision is not a result from a developer mistake (a developer mistake will be classified as a human made fault).

This is where the conventional classification fails, a fault that can be *"software"* and *"non-human-made"* does not exists because the authors considered pieces of software unable/not allowed to take local decisions but agents are defined as *software* pieces making decisions that are *not human-made* decisions.

Note that every fault can, for a point of view, be considered a *human-made* fault since we are considering a human-made system and then agent B would not have such a behavior if the designer/developer made it different. That is not the purpose of the fault classification to make such a judgment on responsibilities linked to the fault (is the human responsible for the acts of the agent?), the only purpose is to characterize the faults. If not, this *human-made* reasoning would have provided less than 25 fault classes in the fault classification, one would have been sufficient.

To sum up the arguments pointed out into this chapter, as the presence of autonomous agents in a system might provide unclassifiable faults and some other researches in the MAS domain defined such faults, we argue that an expansion to the fault classification is necessary.

This expansion is presented in the next chapter.

References

1. Avizienis, A., Laprie, J.-C., Randell, B., Landwehr, C.: Basic concepts and taxonomy of dependable and secure computing. In: IEEE computer society (ed.) IEEE transactions on dependable and secure computing, pp. 11–33 (2004)
2. Arlat, J., Crouzet, Y., Deswarte, Y., Fabre, J.-C., Laprie, J.-C., Powell, D.: Tolérance aux fautes. In: Banâtre (ed.) Encyclopédie de linformatique et des systèmes dinformation Partie 1 La dimension technologique des systèmes dinformation Section 2 Larchitecture et les systèmes, pp. 1–32. Vuibert (2006)
3. Nelson, V.P.: Fault-tolerant computing: fundamental concepts. Computer **B**(7), 19–25 (1990)
4. Klein M., Dellarocas, C.: Exception handling in agent systems. In: Etzioni, O., Müller, J.P., Bradshaw, J.M. (eds.) Proceedings of the Third International Conference on Autonomous Agents (Agents'99), pp. 62–68. ACM Press, Seattle (1999)
5. Platon, E., Sabouret, N., Honiden, S.: A definition of exceptions in agent oriented computing. In: O'Hare, G., O'Grady, M., Dikenelli, O., Ricci, A. (eds.) Engineering Societies in the Agent World'06, 2006
6. Fedoruk, A., Deters, R.: Improving fault-tolerance by replicating agents. In: Proceedings of the First International Joint Conference on Autonomous Agents and Multiagent Systems: Part 2, pp. 737–744. ACM Press, Newyork, USA (2002)
7. Guessoum, Z., Faci, N., Briot, J.-P.: Adaptive replication of large-scale multiagent systems: towards a fault-tolerant multi-agent platform. In: Proceedings of the Fourth International Workshop on Software Engineering for Large-scale Multi-agent Systems, pp. 238–253. Springer-Verlag, Heidelberg (2006)

8. Shah, N., Chao, K.-M., Godwin, N., James, A.E.: Exception diagnosis in open multi-agent systems. In IAT, pp. 483–486 (2005)
9. Chopinaud, C., Seghrouchni, A.E.F., Taillibert, P.: Prevention of harmful behaviors within cognitive and autonomous agents. Paper presented at european conference on artificial intelligence, pp. 205–209 (2006)
10. Dragoni, N., Gaspari, M.: Crash failure detection in asynchronous agent communication languages. Auton. Agent. Multi-Agent Syst. **13**(3), 355–390 (2006)
11. Sabater, J.: Sierra. C.: Review on computational trust and reputation models. Artif. Intell. Rev **24**(1), 33–60 (2005)
12. de Weerdt, M., ter Mors, A. Witteveen, C.: Multi-agent planning: an introduction to planning and coordination. In: Handouts of the European agent summer school, pp. 1–32 (2005)
13. Hägg. S.: A sentinel approach to fault handling in multi-agent systems. In: Second Australian Workshop on Distributed AI in Conjunction with the Fourth Pacific rim International Conference on Artificial Intelligence, pp. 181–195 (1996)

Chapter 4
Refinement of the Fault Classification for MAS

Abstract Considering faults not matching the existing combinations of values presented in the conventional fault classification, as presented in the previous chapter, the classification obviously began by trying to allocate them one of the two existing values of the first classification attribute: the *phase of creation or occurrence*. This attribute is related to the instant when the fault is made, i.e. *development* for faults made during system development (before the delivery of the system to a user) or *operational* for faults made during service delivery (when the system is delivering a service to a user). This approach was finally deemed inefficient as concluded from the examples collected from the Multi-Agent System (MAS) community. Before any fault class corresponding to the values combination obtained at the end of the previous chapter we will answer the following question: May fault resulting from autonomy and pro-activeness be classified as *development* ones or as *operational* ones?

Keywords Multi-agent system · Dependability · Fault classification · Autonomy · Behavioral faults

4.1 The First Attribute Issue

To sum up what we explained in Chaps. 2 and 3, autonomy and pro-activeness enable an agent to have a behavior such as:

- To decide it has a good reason to not respond and to thus never send any answer to a request.
- To voluntarily commit a fault to affect another agent.
- To physically attack another agent, like temporary spam.[1]

---

[1] Note that spam is a kind of physical attack because of the overload of memory, processor or communication means that will result.

K. Potiron et al., *From Fault Classification to Fault Tolerance for Multi-Agent Systems*, 21
SpringerBriefs in Computer Science, DOI: 10.1007/978-1-4471-5046-6_4,
© The Author(s) 2013

We will not be considering for the rest of this chapter the classical faults already present in the fault classification since we seen in Sect. 3.2.1 that this faults are relevant for MAS. Please keep in mind that we are not discussing some developer mistake or disconnection of a cable, we are discussing social faults [1], emerging unwanted behavior [2] and other faults resulting from an autonomous and pro-active agent behavior.

When considering faults resulting from these behaviors for the first time, we tried to classify them with one of the two existing values of the first classification attribute (refer to Chap. 3 or [3] for presentation of the fault classification):

- *"Phase of creation or occurrence"*.

The possible values for this attribute, presented in Table 3.1, are:

- *"Development faults"*, i.e., faults occurring during system development (they occur before the execution of the considered program) that implies that autonomy is a bug;
- *"Operational faults"*, i.e., faults occurring during service delivery during the use phase (they occur when executing the system in question, interacting with programs or human beings).

Assigning one of these values to the faults resulting from an autonomous and pro-active agent behavior means:

- Considering these faults as *development* faults, while autonomy is a *natural* and so acceptable feature of agents (it's an important part of their definition);
- Considering these faults as *operational* faults, whereas they are not, in fact, linked to service delivery but to the autonomous and proactive behavior of the agents and so simply a result of decisions based on the agents perception of their environment, the fault may not be linked with a provided service.

Indeed a fault resulting from an autonomous and pro-active agent behavior is a fault "intentionally" created by an agent (even if the will of the agent is not to make a fault it make the choice of this action) but this fault may not prevent the agent that are at stake to deliver a service. Nor an agent that is not delivering a service will be prevented to make or experience a fault resulting from an autonomous and pro-active decision.

Operational faults are solely and strongly linked to service delivery, which is the reason why faults resulting from the autonomous behavior of agents cannot be considered as operational faults.

We can explain the difference between faults resulting of an autonomous and pro-active agent behavior and *operational* faults when considering that conventional systems are created mainly to provide a service, and are thus judged only with regard to this issue, whereas agents are created to provide a service (and maybe more than one service at a time) as autonomous and proactive entities, and they may be judged with regard to their service delivery but also with regard to

Fig. 4.1 Autonomy
integration into fault
classification values

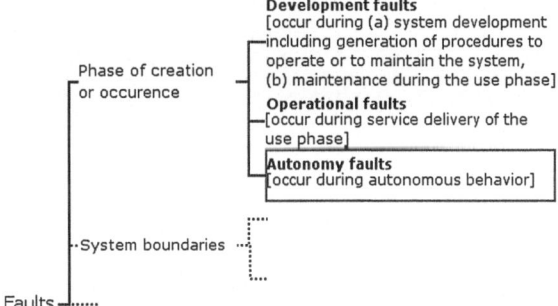

their behavior and the fact that this behavior is a result of decisions based on the agents perception of their environment.

For this reason, we propose to introduce *autonomy* as a new value of the attribute *"Phase of creation or occurrence"*, as it can be seen on Fig. 4.1. The autonomy value will represent faults occurring during the "autonomous and pro-active behavior" of an agent. By "autonomous behavior", we mean all actions that autonomy allows the agents, e.g.:

- Not responding to a request (the agent "power to say no") or responding negatively whether or not it is included into the interaction protocol;
- Creating a fault in order to incapacitate another agent;
- Breaking a commitment.

Introducing a new value for an attribute creates 96 new possible combinations; among them, the analysis below shows that 12 correspond to relevant new fault classes.

As fault perception is a matter of perspective, we will present these new fault classes using two different points of view: agent-centered developed in Sect. 4.2 and "external to the MAS"-centered (user-centered) developed in Sect. 4.3.

4.2 Agent Centered Analysis

A fault resulting from autonomous and proactive behavior, from an agent-centered point of view, is equivalent to the "freedom" that autonomy gives to other agents, and their unpredictability (e.g., functions and variables of an agent are not known by the others and each agent has a limited perception of its environment). If an agent displays autonomy, this is not a fault from this agent perspective; the act in question is a fault only for the agent(s) it is interacting with.

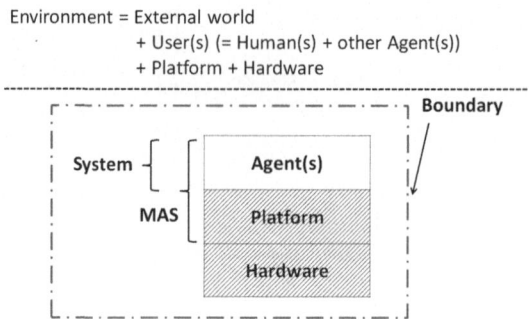

Fig. 4.2   Definition of the considered system as an agent

We define more precisely what is studied here as illustrated by Fig. 4.2:

- **System/point of view**: one agent;
- **Environment**: other agents and/or human being;
- **Boundary**: agent communication system;
- **User(s)**: other agents and/or human being.

We give six situations to illustrate our purpose.

1. An agent from time to time voluntarily commits a fault to interfere with another agent. For instance, sending a wrong message because it has chosen not to follow a correct interaction protocol (if the protocol known by the agent were not correct it would be a development fault).

   This is a useful behavior to steal resources to other agent(s): look there you will find what you want, and during this time I will be alone here.

2. A voluntarily committed fault, as in example 1, but permanent.
3. An agent estimates that it has no time to respond and so the other agent does not receive any answer (the duration of such a fault may be time-bounded and is directly linked to the agent context).

   This is a useful behavior when the agent is overloaded: I do not respond and only the important requests will be re-emitted.

4. An involuntarily fault as in example 3, but not bounded in time.
5. Physical attacks between agents, like temporary spam.

   This is a useful behavior when it is needed to eliminate a concurrent: we will have to manage what I sent you before answering to this auction; I will have time to win it.

6. Idem physical attacks as in e.g., 5, but permanent.

For these faults, the values of the attributes are:

- *Phase of creation or occurrence*:

  - *Autonomy* (by definition)

- *System boundaries*:

  - *External*; because its source is the other agent (an "internal to the agent" fault would be a development fault).

- *Dimension*:

  - *Software*; autonomy comes from the agent's implementation (examples 1–4);
  - *Hardware*; autonomy does not come from the hardware but affects it (examples 5 and 6).

- *Phenomenological cause*:

  - *Natural*; autonomy does not allow a human being to dictate its behavior to the agent (by definition).

- *Objective*:

  - *Non-malicious*; (examples 3 and 4);
  - *Malicious*; (examples 1, 2, 5 and 6).

- *Capability*:

  - *Deliberate*; results from the decision of an agent.

- *Persistence*:

  - *Transient*; if the decision context is bounded in time (examples 1, 3 and 5);
  - *Permanent*; (examples 2, 4 and 6).

This classification is represented by the tree of fault classes number 32–37 on Fig. 4.3.

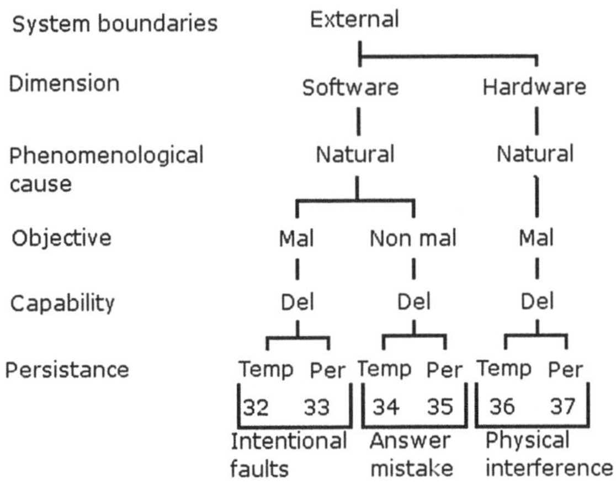

Fig. 4.3 External behavioral faults

System boundaries	External	
Dimension	Software	Hardware
Phenomenological cause	Natural	Natural
Objective	Mal Non mal	Mal
Capability	Del Del	Del
Persistance	Temp Per Temp Per	Temp Per
	32 33 34 35	36 37
	Intentional Answer	Physical
	faults mistake	interference

Except for the values *natural* and *software* combination that we discussed before (Sect. 3.2.2) and the use of the *autonomy* value, these six fault classes does not introduce a new value combination.

4.3 System Centered Analysis

A behavioral fault in the "external to the MAS"-centered (or user-centered) point of view is comparable to the incompetence of the MAS or of one or more agents to handle the autonomy of some agents. This refers to how an agent can handle the autonomous behavior of the agents it interacts with.

From an external point of view, the faults can be caused by one agent with malicious intent and are perceived at user level only because of the system incompetence/incapacity to handle the fault.

We define more precisely what is studied here as illustrated by Fig. 4.4:

- **System/point of view**: a MAS;
- **Environment**: other agents, other MAS and/or human being;
- **Boundary**: other MAS communication system;
- **User(s)**: other agents, MAS and/or human being.

We give six situations to illustrate our purpose.

1. An agent overloads the network creating temporary problems due to message transmission durations.
2. Physical fault as in example 1 but not bounded in time.
3. An agent is incompetent to achieve its goal because of another agent reaction (request refusal) and temporarily has no other way to achieve its goal.
4. Incompetence fault as in example 3, but permanent.
5. An agent voluntarily creates a temporary fault to prevent another agent from accomplishing its goal.
6. Malicious fault as in example 5, but permanent.

Fig. 4.4 Definition of the considered system as a MAS

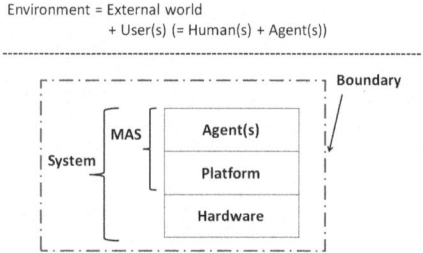

For these faults the values of the attributes are:

- **Phase of creation or occurrence**:

  - *Autonomy*

- **System boundaries**:

  - *Internal*; because its source is in the MAS.

- **Dimension**:

  - *Software*; autonomy comes from the agent's implementation (examples 3–6);
  - *Hardware*; autonomy does not come from the hardware but affects it (examples1 and 2).

- **Phenomenological cause**:

  - *Natural*; autonomy does not allow a human being to dictate its behavior to the agent.

- **Objective**:

  - *Non-malicious*; (examples 3 and 4);
  - *Malicious*; (examples 1, 2, 5 and 6).

- **Capability**:

  - *Incompetence*; results from an agent inability to adapt to the unexpected behavior of other agents or to changes in the environment.

- **Persistence**:

  - *Transient*; if the decision context is bounded in time (examples 1, 3 and 5);
  - *Permanent*; (examples 2, 4 and 6).

This classification is represented by the tree of fault classes number 26–32 on Fig. 4.5.

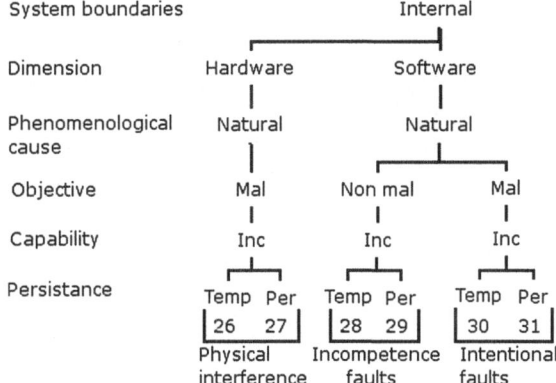

Fig. 4.5 Internal behavioral faults

Faults from this "external to the MAS" point of view are observable only if an agent is incompetent to handle some others autonomous behaviors and the MAS is incompetent to tolerate this agents incompetence. This explains the introduction, with the new faults, of new combinations of values for the *objective* and *capability* attributes.

4.4 Faults Review

As shown at the bottom of Fig. 4.3 and Fig. 4.5 (and summarized in the last line of Fig. 4.6 and Table 4.1), several agents faults can be grouped into examples of fault groups.

- The group of *malicious software* faults is called intentional fault group, since they are faults committed intentionally by agents.
- The group of *external non-malicious deliberate* faults is called answer mistake, since they are committed with no bad intentions.
- The group of *internal non-malicious incompetence* faults is called incompetence fault group, since they result from the agent's incompetence with regard to other agents autonomy.
- The *hardware fault* group extends the physical interference group, as they are close to this example class.

Concerning the three non-exclusive fault groups (development faults, physical faults or interaction faults) presented on Table 4.1, the faults whose first attribute is "autonomy" cannot be considered as belonging exclusively to a group. The

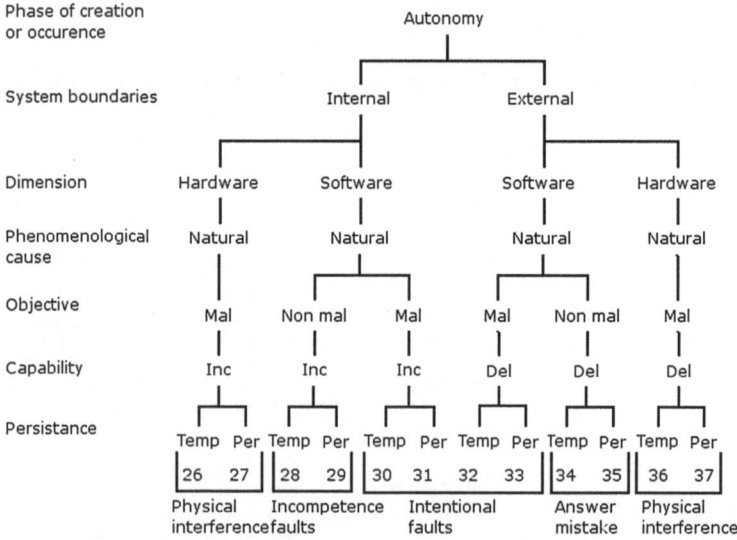

Fig. 4.6   Behavioral faults

Table 4.1 Behavioral faults example classes

Behavioral faults											
Development faults								-			
-								Interaction faults			
Physical faults		-								Physical faults	
26	27	28	29	30	31	32	33	34	35	36	36
Physical interference		Incapability faults		Intentional faults				Answer mistake		Physical interference	

"autonomy" faults include faults like intentional spam (which belongs to the physical fault group), response defaults (belonging to the interaction fault group) or wrong response (belonging to the development fault group). We therefore named these faults *behavioral faults*, and consider it a fourth non-exclusive group of faults. It is shown in the first line of Table 4.1.

Nevertheless, some behavioral faults can be classified in the same time as development, physical or interaction faults, as shown on Table 4.1.

Faults 26–32 are *development* faults, since these faults originate in the agent or system incompetence to handle agents autonomous behavior. They can cause service outage and force the system into a degraded mode or, at worst, stop execution.

Faults 32–37 are *interaction* faults because all are external to the system considered. They can cause local service failures (not always observable from an external viewpoint).

Several ones of faults can also be viewed as *physical* faults (faults 26, 27, 36 and 37) because of their influence on hardware.

4.5 Validity of This Approach

The need for the classification refinement may not be clear enough at this point of the argumentation. The next sections analyze the differences between the conventional faults and the behavioral faults and then discuss the reasons of these differences.

4.5.1 Faults Comparison

In order to analyze the relevance of the behavioral faults, we made a comparison to evaluate their distance with regard to pre-existing faults. This comparison is based on a computed Hamming distance [4] representing the number of different values between the attributes describing two faults.

For example, the Hamming distance for faults 20 and 32:

- Fault 20 is described as:

  - "**Operational**, External, Software, **Human-made**, Malicious, Deliberate, Temporary".

- Fault 32 is described as:

  - "**Autonomy**, External, Software, **Natural**, Malicious, Deliberate, Temporary".

Their Hamming distance is equal to 2.

This example and others are shown on Fig. 4.7, red circles states for conventional fault classes attributes having the same value that fault class 32 so that to calculate the Hamming distance from two fault classes is to count the fault class attributes without red cycles.

The highest possible Hamming distance is *seven* because the fault classification has seven attributes; moreover the first attribute ("phase of creation or occurrence") always has a different value for conventional faults and behavioral faults so the distance cannot be below *one*.

We then count the hamming distance of all behavioral faults with regard to conventional faults so that to be able to discuss the relationship between them. Results are presented on Table 4.2 where rows are behavioral faults (BF), columns are Hamming distance (HD) and the boxes are filed with the number corresponding to conventional faults having the corresponding Hamming distance with the behavioral faults. Or to put it differently, a cell in row i and column j contains all the conventional faults at Hamming distance j with behavioral fault number i.

This table shows that some conventional faults are very similar to behavioral faults. But the first observation is that there is no Hamming distance below 2, that is to say that behavioral fault classes have at least 2 attribute values differing from conventional fault attributes combination.

This difference had first been pointed out when analyzing the reported faults in MAS (Sect. 3.2.2), which had underlined the *software* and *non human-made* combination being representative of faults experienced by MAS but not being included in the conventional faults classification.

It tends to confirm the need for introducing behavioral faults, since they are not redundant which would have been the conclusion if the Hamming distance were 1 because it would have proven that only the *phase of creation or occurrence* and the introduced *autonomy* value would have deferred and the behavioral faults would be only a transposition of conventional faults.

Moreover, this table also allows noticing that no faults have a distance of 5, 6 or 7 what would have shown that the behavioral faults were very different from conventional faults, and this difference would have possibly denote imaginative values combinations more than real ones. The logic sustaining the conventional faults definition is not broken by the behavioral fault introduction.

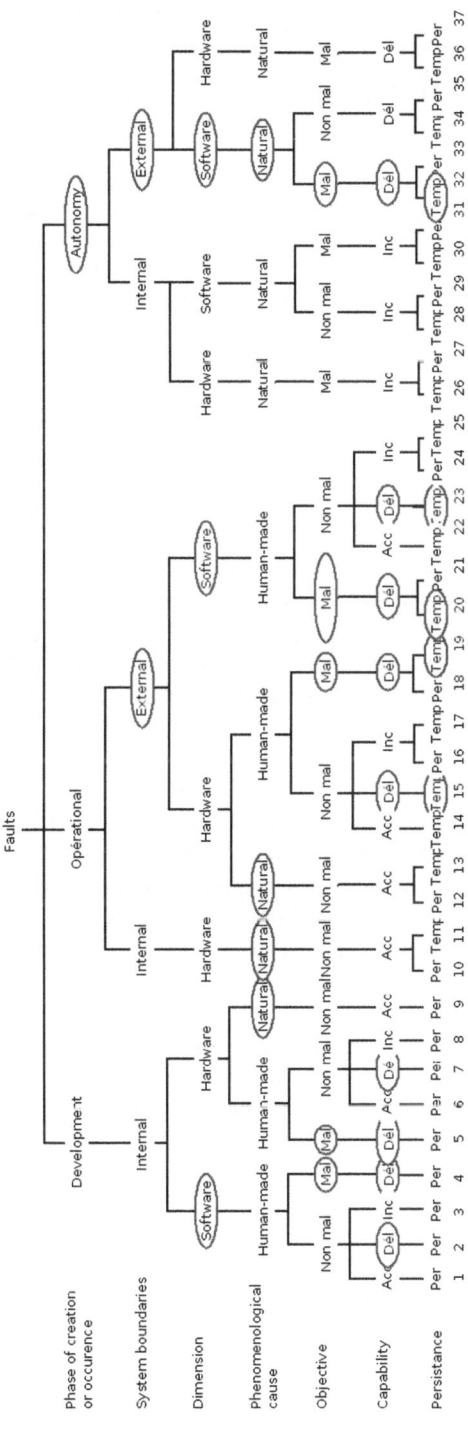

Fig. 4.7 Comparison of fault 32 to conventional faults

Table 4.2 Faults comparison

BF \ HD	4	3	2
26	5, 8, 9, 13, 17, 19	11	
27	3, 4, 6, 7, 12, 16, 18	5, 8, 9, 10	
28	8, 9, 13, 17, 22, 23	3, 11, 25	
29	4, 6, 7, 12, 16	1, 2, 8, 9, 10, 24	3
30	3, 4, 11, 20, 25		
31	1, 2, 5, 8, 9, 10, 21, 24	3, 4	
32	4, 13, 15, 18, 22, 25	19, 21, 23	20
33	2, 5, 12, 19, 23, 24	4, 18, 20	21
34	2, 11, 12, 14, 17, 19, 21, 24	13, 15, 22, 25	23
35	1, 3, 4, 7, 9, 10, 13, 18, 22, 25	2, 12, 16, 21, 23, 24	
36	5, 11, 12, 14, 17, 21, 23	13, 15, 20	19
37	4, 7, 9, 10, 13, 15, 16, 20	5, 12, 21	18

Another observation is that fault number 3 (a development fault) is very present in this table. It is not possible to draw conclusions at present, but some questions and observations can be raised.

Does this observation on fault 3 mean that, in terms of software development, "autonomy" is close to the capability to violate requirements (characteristic of fault 3)?

In terms of software development, autonomy is a requirement of the agents so that considering that autonomy is a requirement that allows agent to violate the rest of the specification as no sense since all requirements should be consistent. But considering of the effect of such a behavior for the system and difficulty for system designers and developers considering that other system entities should be able to be autonomous or to violate a requirement is quite the same.

Another comment about this observation is that in terms of system dependability the violations of protocols and requirements are usually the work of fault prevention, whereas for agents these violations may be tolerated when the system is working because autonomy cannot be "corrected" as it's a part of the agent definition. This fact is another argument emphasizing the importance of fault tolerance for MAS.

Moreover, does the fact that this fault 3 appears five times imply that MAS are more subject to some kinds of fault than conventional distributed systems?

To answer this observation, we have now to notice a particularity of the fault classification that is that it takes no consideration on the probability that the fault may occur so such a consideration that is stated in the question must not be made. In some MAS agents will not be using malicious behavior because the system is not based on competition so their autonomy manifestations should be only protective behaviors and so quite unnoticeable and in other MAS where competition is the law the agents autonomy should cause much more faults. But ideationally MAS are not designed and agents were not defined to be a never-ending source of faults.

The fault class number 30 (*autonomy, internal, software, natural, malicious, incompetence, permanent*) is quite different from all the conventional faults. This fault is related to the incapability of the MAS to tolerate the autonomous behavior of the agents. This is the break point between fault tolerance for MAS and for classical systems: MAS cannot be of any use if these faults are not handled and MAS design at system level cannot ignore the behavioral faults. Whereas we explained when presenting the faults classification that fault classes that should be tolerated may be chosen into the fault classification there is no choice for MAS regarding those faults. However, these faults (*internal* behavioral faults, fault classes 26–31) are the farthest ones in the comparison between behavioral faults and conventional faults what leads us to consider that fault handlers suitable of conventional systems will not be adapted for MAS. Agents and MAS need suitable/adapted methods and mechanisms to be fault tolerant.

Note that whereas it can be considered that conventional systems should not be adapted for "autonomous behaviors" because autonomy is not part of their definition and so behavioral faults are not considered to be experienced by the system and components.

The last observation we want to underline is the closeness of conventional faults and behavioral faults with regard to interaction faults that is the result of the closeness of human-being interaction faults and agent interaction faults (human is a really autonomous agent to be integrating into a system). This put into perspective that some conventional fault tolerance methods defined for human being interactions can be used, or at least adapted to MAS in order to handle groups of interaction faults belonging to conventional faults and behavioral faults. These adapted methods could be able to handle interaction faults due to autonomy as well as those due to operational context. But it also points that tolerating behavioral faults will be difficult because interaction faults are faults that came with less information and lot of diagnosis difficulty.

4.5.2 Analysis of the Difference Between Faults

Our Hamming distance study underlines some similarities between conventional faults and behavioral faults, but also many differences we explain below.

For *natural* faults.

Every faults introduced in our analysis are *natural* faults, since autonomy is a natural component of the agents. The difference is that for conventional faults, the phenomenological cause being *natural* implies the combinations of values to only verify *non-malicious* and *physical* values but when in agent systems it can be both *natural*, *software* and *malicious*.

This is partly explained by taking into account the fact that agents are programs allowed to be malicious. They can act not only at a physical level, but also at the interaction level which is the new combination introduced from the study of MAS faults made in Sect. 3.2.2 (*non-malicious* and *software*). The consideration we made on autonomy would tend to unify human faults and faults made by agents because *software* faults were for conventional faults only *human-made* faults.

For *malicious* faults.

Malicious faults are no longer only *human-made*. Autonomy brings into the MAS entities that have specific features usually reserved for humans (*autonomy* and *pro-activeness*). In particular in open MAS, agents are not always cooperative: they can decide to carry out malicious actions towards other agents.

An interesting point to stress here is behavioral faults prevention. A simple way to prevent all non-malicious behavioral faults is to send a preventive message. For instance, if an agent has too many messages to consider, it can send (without reading the messages, to save time) a cancel message, or if it cannot deliver a result in time, ask for a delay before receiving the reminder.

Prevention messages do not increase the amount of exchanged messages, since they are only used for exceptional situations and since a fault-handling mechanism is required anyway (sending a preventive message is as much message for the MAS than sending a reminder). However, this is not entirely suitable to prevent malicious faults, because prevention will be useless if a fault is made with malicious intent: there is no way to dictate its behavior to an agent or to know precisely how to influence its decisions so there is no way to change its "malicious way of thinking".

For *interaction* faults.

Half of behavioral faults are *interaction* faults. They are very close to *interaction* conventional faults; one outstanding point is that they differ according to their classification as *natural*. The similarity between conventional faults and behavioral faults leads us to the conclusion that, at least for interaction faults, some of conventional faults tolerance methods can be used, or adapted, to MAS in order to handle groups of interaction faults belonging to conventional faults and behavioral faults. These adapted methods would be able to handle interaction faults due to autonomy as well as those due to operational context.

All of these observations lead us to the consideration that faults related to agent autonomy can be treated partly as human interaction faults, and partly as development or physical faults, because their effects will be very similar.

4.6 Preliminary Conclusion

We have seen first that autonomy and pro-activeness of the agents are sources of faults.

But the fault comparison and connection we carried out between conventional faults and behavioral faults raises the hope that tolerance to them will naturally make agents tolerant to other classes of faults—particularly development and interaction faults. The processing needed to tolerate these faults could thus be factorized. Since behavioral faults might naturally happen in MAS applications, MAS applications have to be behavioral faults-tolerant, they might incidentally be more robust to some other kinds of faults, the "nearest" kinds of faults what strives to be an argument for the MAS robustness argument of MAS community viewed in Sect. 2.2.

Plus, from a study we made on time redundancy [5], it seems that contrary to what would be expected, fault tolerance methods used for conventional systems cannot guarantee the handling of behavioral faults because of the fundamental difference of assumptions made when interacting with autonomous agents. But such methods can be adapted to handle the behavioral faults without losing the handling possibilities for conventional faults [6]; this will be explained in the next chapter.

Since the aim of this book is to study dependability, and in particular specifications needed for MAS fault tolerance, the next section will present a specific fault handler made for Multi-Agent Systems faults.

References

1. Shah, N., Chao, K.-M., Godwin, N., James, A.E.: Exception diagnosis in open multi-agent systems. In: IAT, pp. 483–486 (2005)
2. Fedoruk, A., Deters, R.: Improving fault-tolerance by replicating agents. In: Proceedings of the First International Joint Conference on Autonomous Agents and Multiagent Systems: Part 2, Bologna, Italy, pp. 737–744. ACM Press, New York, (2002)
3. Avizienis, A., Laprie, J.-C., Randell, B., Landwehr, C.: Basic concepts and taxonomy of dependable and secure computing. IEEE Trans. Dependable Secure Comput. 1, 11–33 (2004)
4. Hamming, R.W.: Error detecting and error correcting codes. Bell Syst. Tech. J. 26(2), 147–160 (1950)
5. Potiron, K., Taillibert, P., Fallah-Seghrouchni, A.E.: A new performative for handling lack of answers of autonomous agents. In: ICAART, pp. 441–446 (2009)
6. Potiron, K., Taillibert, P., Fallah-Seghrouchni, A.E.: A step towards fault tolerance for multi-agent systems. In: Languages, Methodologies and Development Tools for Multi-Agent Systems First International Workshop, September 2007. Revised Selected Books Lecture Notes in Computer Science, Vol. 5118, pp. 4–6 (2007)

Chapter 5
Fault Tolerance for MAS Specific Faults

Abstract An agent can send a message but never receive any response, which leads to handling problems. This issue, called the "empty mailbox problem", is addressed in this chapter. The causes of this problem can lie in MAS low level layers, for instance in communication links, but also in the higher layers, for instance in the behavior of autonomous agents, which can independently choose not to respond. It is not an easy task for the designer/developer of the agent to determine what is to be done in such cases and how to do it. The proposed solution for handling this empty mailbox problem consists in introducing a new performative and its associated protocol, producing a generic way to handle the empty mailbox problem in the case of transient faults.

Keywords Multi-agent systems · Agent architecture · Fault tolerance · Resend · Autonomous agent · Adaptability · Performative · Behavioral faults

5.1 Context

The Empty Mailbox Problem (EMP) represents the problem of not experiencing the expected event after doing an action, for instance: receiving no response after sending a message that requires one. In this case, the agent designer/architect addresses the issue of deciding what is to be done when the program does not perceived an expected piece of information. Since this issue can apply to all messages the agents send and all other actions it can do, a generic way to handle it could be very useful.

This issue is not a new one nor is it specific to Multi-Agent Systems (MAS): in distributed systems, such a lack of expected perception (the fact that an expected message is not received) is also addressed. There are some specific handlers used according to the possible cause, for instance a development fault (improper interaction protocol implementation), communication link malfunction (message lost or altered), hardware failure (computer crash), improper interaction protocol specification or malicious user.

But for MAS composed of autonomous agents, another cause can be identified: autonomy and so the behavioral faults defined by the Chap. 4. The behavior of an autonomous agent may differ from the expectations of the other agents and so an autonomous agent may choose, for whatever reason, not to follow the interaction protocol used in a conversation with another agent, so the latter agent mailbox will stay empty whereas it is expecting a message.

A more rigorous definition of EMP is: an error that agents communicating by passing messages can experience due to certain fault, and that can cause deadlock. Since faults are defined as the cause of errors, a precise definition of those causing the EMP is needed to complete the study in order to obtain clues indicating the appropriate handler. This is done in the next section, through analysis of the fault classification results obtained so far.

5.1.1 Faults Leading to EMP

The fault classes defined by the fault classification of Avizienis et al. [1] presented in Chap. 3 are, in particular, applicable to distributed systems (which MAS are). A study of the faults leading to EMP in distributed systems reveals that they can potentially belong to any given class. These include development (a mistake in the code of the agent), interaction (an incorrect specification of the interaction protocol) or operational (an unplugged network cable) faults.

A study of the faults leading to EMP in MAS reveals that on top of being from any classical fault class, they can potentially belong to any of the *behavioral* fault classes.

Note that from this point of view the fault classification should look useless and not enough discriminative. The root cause analysis made to find clues on the faults leading to the EMP was easier using defined and limited number of fault classes than all possible faults. The synthetic view of the diversity of the faults and the fault classes used as a check-list for the study of the possible faults where the better possibility we found for addressing faults issues.

EMP is an error resulting from any of the possible fault classes occurring in MAS. This shows the advantages of reducing agent interactions only to message-passing with a view to isolating the fault, errors resulting from many of the faults affect mostly the agent means of communication (even if other errors may occur) the more interaction means the agent has the more protection will be needed.

5.1.2 Consequences

The first consequence deduced from the fault study is that defining a handler adapted to a specific class of faults or a specific interaction protocol may result into a mass of handlers. Given that message passing is an important means of communication for agents (if not the only one), it quickly becomes a critical point in the event of failure and its treatment becomes mandatory. Thus, making handlers adapted to a specific situation (for a specific agent state, interaction protocol or specific system) or for a specific class of faults cannot lead to an easy solution to EMP but rather increase the difficulty for agent designer/developers to find an appropriate handler.

Furthermore, all faults cannot be detected during the system design and testing since many result from unexpected situations (e.g. a power cut resulting from cleaner needing to plug a vacuum cleaner), result from unpredicted faults combinations (i.e. fault easily handled when alone but critical when occurring simultaneously) or have a byzantine behavior. This is particularly true when considering behavioral faults. These faults are linked to agent autonomy and pro-activeness: they cannot be eliminated because they are intrinsic to the nature of the agents, so correcting them is irrelevant. They can, at most, be controlled or tolerated. This explains why our work concentrates on fault tolerance and not on other means of achieving dependability, such as fault elimination, fault prevention or fault forecasting. Moreover, we consider inevitable to introduce fault tolerance into MAS because of these reasons.

5.2 Related Work

Part of the EMP had been addressed in the distributed system domain, which gives some possible ways of handling it. This related work will be presented with regard to its sustaining time models.

Acknowledgments can be used for asynchronous message passing when the number of sent messages is not restricted. But this may be useless for MAS, since the asynchronous model cannot be used to implement consensus, election or membership [2]. Acknowledgments might not persuade an autonomous agent to answer a request; indeed they do not address some behavioral faults.

The synchronous model specifies that any message sent using the correct process to a correct destination is received [2], but such assumptions are quite difficult to guarantee. Furthermore, the model offers no solution to EMP because autonomous agents can be correct yet not send any message.

There are also intermediate models, like the timed asynchronous one [2]. This model allows correct processes not to send a message and can solve consensus problems on some assumptions. In this case, one way of handling communication issues is to use time redundancy [3] if the fault is assumed to be temporary. Time redundancy (or retry) corresponds, for communication issues, to sending the same message again and again during a fixed period while the expected response is not received. Such a method is not adapted to autonomous agents; because it might not persuade an autonomous agent to answer (it does not address some behavioral faults). The following example illustrates this fact: Consider an agent (buyer) responsible for buying books. The buyer does some research, finds the lowest price for the book and contacts the agent selling the book (seller) in order to pay for it. After sending the payment message, something goes wrong and the buyer's mailbox remains empty. If the buyer uses the retry method it will send the payment message numerous times, with the risk it entails (i.e. multiple purchases).

If time redundancy does not provide any response then rollback or rollforward [1] is used, the software will respectively return to its last saved safe state or define a new safe state.

Moreover, since the retry method can be confused with a repetition of the same message possibly due to a development or a malicious fault, the agents might ignore the repeated messages considering that they come from a malicious or faulty agent. In the previous example, being seen as malicious may result in receiving no book and never being allowed to buy another book from this agent.

In the MAS domain, two main lines of research can be highlighted that include fault tolerance in or outside the implementation language of the agents.

Examples of the first line, Cool [4] and AgenTalk [5], present some agent communication languages including prospects about mechanisms for handling faults. They propose using a meta-script to control the execution of behavioral scripts. The meta-scripts start with messages sent by the behavioral script representing exceptions. The authors do not provide any general control meta-script, and assume that they will be constructed suitably for each exception. They conclude predicting that meta-scripts can be grouped together using exception classes.

Similarly, 2APL [6] (A Practical Agent Programming Language) takes into consideration that when an agent performs an external action, this action might fail. To do so, it uses the time redundancy handler presented before. A time-out parameter is used for specifying a time-span in which actions can be retried again and again in case of failure. While the execution of the external action fails and the time specified by the time-out has not elapsed the action is executed again. In this case the agent is oblivious about the action failure. Since the action (and time redundancy) might fail, an external action may be eligible for repair by a PR–rule (Plan Repair rule).

Another approach is developed in FT-ACL [7] (Fault Tolerant Agent Communication Langage): the authors propose a fault tolerant agent communication language which deals with agent crash failures, providing fault-tolerant communication primitives and support for an anonymous interaction protocol. The knowledge level of the FT–ACL is based on giving, for each act of speech, a success continuation and a failure continuation, but no general treatment is provided for the failure continuation.

These methods are related to conversational agents and the handling of EMP, faults perceived in the agents communications. Only 2APL provides the agent developer with a fault handler, but for message passing this handler has some issues, as explained before. In the other languages and in the event of action retry failure in 2APL, defining suitable fault handlers is the designer's/developer's responsibility and task. A serious problem with such a handler creation is the risk for the designer/developer of introducing new faults into handlers.

5.3 A Handler for the Agents

5.3.1 Analysis of the EMP

The first observation that must be taken into consideration is that agents have a limited perception of their environment, and it is thus impossible for them to diagnose the fault. Moreover, even if agents could diagnose that the fault is due to the expression of autonomy, they have no power to enforce obedience in others. So the proposed handler must help with most of the fault classes and be particularly adapted to autonomy.

But the proper handling of a fault depends directly on whether the fault is temporary or permanent. Indeed, methods adapted to the first case are not relevant to the second, and methods suitable for permanent faults are too costly or too drastic to use for temporary fault. This section focuses on temporary faults, and the proposed handler usefulness of the define handler with regard to these faults and permanent ones is discussed in Sect. 5.4. Focusing on temporary faults is a restrictive working assumption, but addresses half of the fault classes. Furthermore, since temporary faults are difficult to find during code verification and tests, they can be more frequent than permanent ones during running.

This is the only restriction made concerning the handled faults' characteristics, since seeking a generic method facilitates the development of fault tolerant agents, whereas building specific handlers implies more work for the designer, as well as being a new source of faults.

However, we consider that a handler incorporated into the behavior of the agent may facilitate detection of the fault's consequences for the agent's developer. Indeed, a fault in a conversation may have many consequences on the agent's entire behavior. For instance, a delay in obtaining a response can delay the reply to another agent. Furthermore, incorporating a handler in the agent's behavior allows it to be used with any of the methods existing in the MAS domain. For example it may be integrated in the control meta-script of Cool or AgenTalk, in FT-ACL failure continuation, or proposed as a handler for the sentinels.

We assume that agents are autonomous, and therefore free to assess whether the fact that the expected message has not yet arrived is a "normal" or an "exceptional" situation, i.e. capable of detecting EMP. We assume that in every state corresponding to waiting for a specific response or event, the agent has fault detection means (e.g. a timeout or a comparison between the observed behavior and the expected one). This serves to avoid agents waiting endlessly for a message, and to detect a fault before any other action is carried out locally (at the considered conversation level).

Moreover, we consider that the agent must have the opportunity to assess fault handler usefulness with regard to its current internal state. Thus agents must have handlers adapted to their knowledge level, and some knowledge of how to choose and use them and we will provide that information in the next sections.

5.3.2 Solution Key Idea

The key idea as presented in [13] can be illustrated as follows:

- When your web browser tells you something like "The server is down, please try again later", you refresh the page; in doing so you use the retry method (explained Sect. 5.2): sending exactly the same request but when you are on a payment page, this is not recommended because it may result in multiple payments;
- When you e-mail somebody and receive no reply you prefer to send another mail saying "I had no reply to my previous e-mail and that really bothers me"; in doing so, you use another handler: sending a new message to express that you have received no response. If you were able to do so with your browser, you would not have any multiple payment problems.

This second behavior is the key idea of our handler: an acknowledgment of no-reception. It is possible to define such a handler because we are dealing with intelligent agents and not a classical web application.

5.3.3 Formal Description

The acknowledgement of no-reception is named "resend". It is defined using a performative and associated protocol first presented in [8]. It was designed for agents to handle some faults based on the argument that a time redundancy (or retry) can be used cooperatively and with added information to improve its effectiveness and the range of tolerated fault classes.

The agent can obtain some useful information at its knowledge level using a method that is quite similar to a retry: When an agent (sender) thinks that it should have received a response to a message sent previously, it can send another message encapsulating the previous one into a dedicated performative to explain the issue to the other agent (receiver). The message encapsulation makes our method different from a retry and avoids confusion with a stutter (repetition of the same message, potentially due to a development fault or to a malicious behavior).

The handler is then used when an agent deems that it should have received a response to a message (or (not) perceived an expected result after sending a message) to express to another agent that it considers that something went wrong in the conversation. In particular, this handler differs semantically from a time redundancy (sending the same message) and introduces specific fault-handling states into the agent behavior.

The performative is named *resend* and it expresses *a mental state to the other agent*, and so corresponds to an *expressive speech act* as defined in [9].

Being an expressive speech act makes this performative semantically different from FIPA-ACL or KQML, which only know assertive and directive speech acts.

Table 5.1 Resend description for FIPA-ACL formal logic

Name	Resend
Description	An agent i tells an agent j that i wants j to process the expression Φ because i has not perceived any realization
Message	Expression Φ corresponds to the previous message
Semantic	$<i; resend(j,\Phi) > \equiv < i, inform(j,Ui{-}^{\wedge} Ii\Phi)>$ FP : Ii Φ $^{\wedge}$ Ui(BjIi Φ v BjΦ) RE : BjIi Φ

Table 5.2 Resend description for KQML

Name	Resend(A,B,Id)
Semantic	A states to B that it has not received a response to the KQML message identified by Id NOT(know(A,process(B,Id))) Pre(A):NOT(know(A,process(B,Id))) Pre(B):NONE Post(A):know(A,know(B,NOT(know(A,process(B,Id))))) Post(B):know(B,know(A,NOT(know(A,process(B,Id)))))
Content	Performative_name is the performative previously sent that require a response

Formal descriptions of the performative are given using FIPA-ACL [10] in Table 5.1 and KQML [11] in Table 5.2.

The associated protocol is defined as follows and depicted Fig. 5.1:

After detecting a fault, the agent sends a message using the resend performative and waits for a reply in a new waiting state.

Note that this state must also be guarded with a detection mean that will raise a new alarm if no response is received indicating the agent that the fault have not be handled and may not be temporary (or at least, the duration of the fault is exceeding the time the agent choose to wait).

5.3.4 Conversation Formalism

In most cases a fault will be detected because of a lack of message reception but in some cases, a specific perception can signal a fault, so a comparison between the observed behavior and the expected one will be better than a timeout for fault detection. For example, if the agent wants to check that a database is unavailable using a "ping" message, if it receives a "pong" message then the agent will assume there is a fault. In this case, since the agent will wait "a certain time" before considering that the database is unavailable the "timeout" will be the nominal expected continuation of the behavior and the message reception will be the fault detection.

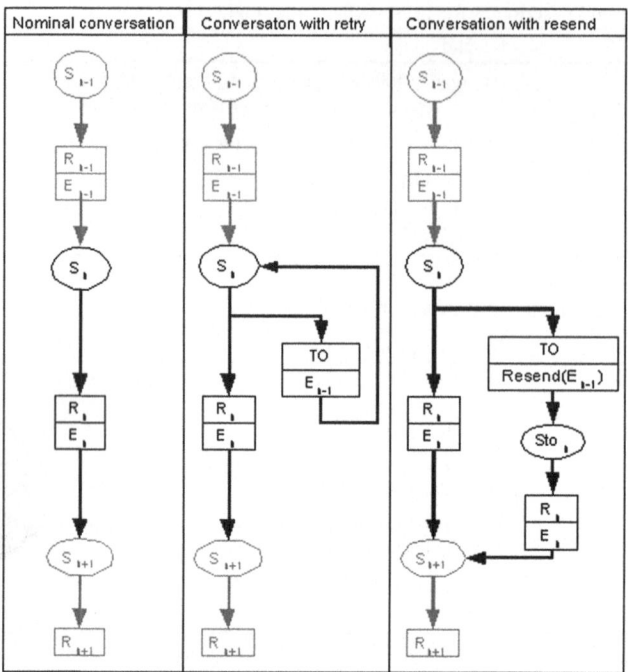

Fig. 5.1 Conversation comparison

Because the timeout is more often representing the "fault detection" than all other perceptions, in the following sections, the term timeout may be used to signify "all detection means".

It must be noted that we use the term timeout as "the maximum waiting time before the agent considers that it is not normal to have received no expected perception (message)" (contrary to 2APL that uses timeout as the maximum time for time redundancy). How to determine the timeout duration, for example through negotiation, specifications or calculation, is not discussed here.

The conversation formalism is intended to make it easier to explain and integrate the proposed solution into interaction protocols. It helps interleave the proposed solution into a conversation level (corresponding to an architectural description of the agent conversation) to increase overall comprehension before systematically deducing agent automatons. It was used to design agents (at architectural level) for our experimentations.

Examples of conversations are illustrated Fig. 5.1: circles stand for conversation states, rectangles both for received or fault detection means (up) and sent messages (down) and arrows for transitions. With the following notations:

- S_n states corresponding to the waiting of m messages with $m \leq \text{card}(R_n)$ and messages $\in R_n$,

- R_n the set of possible incoming messages expected by the agent in the state S_n, they are grouped by subsets (R_n is $\cup r_{nx}$) in order to model conversations in which a group of messages have to be received before sending a response,
- E_n the set of nominal messages possibly sent in the state S_n as a response to R_n, they are grouped by subsets ((E_n is $\cup e_{nx}$)) in order to model conversations in which several messages have to be sent before waiting for a response, e_{nx} is either the beginning of the conversation or the response adapted to r_{nx},
- TO the timeout, maximum waiting time of a subset of needed message in a particular state, used for the example as detection means for the EMP (not that another detection mean easily described by the conversation formalism is the reception of a message $\notin R_n$),
- Leaving a waiting state S_n occurs when a subset of messages $r_{nx} \in R_n$ or a fault detection occurs (a timeout is received). It leads to the end or continuation of the conversation and/or the sent of a subset of nominal messages $e_{nx} \in E_n$.
- Sto_n state corresponds to the waiting state following a fault detection in state S_n,
- The message resend(r_{nx}), the resend performative with $r_{nx} \subseteq R_n$, represent the last set of messages that the agent can receive because they are with no responses for the other agent(s) participating to the conversation,
- The message resend(e_{nx}), the resend performative with $e_{nx} \subseteq E_n$, represents the set of messages with no responses for the agent, it is the message it will send after a fault detection,
- End represents the end of the conversation and Fail represents the failure of the proposed handler (the fail state is not always used since it only represents a particular case of the end).

In particular, this abstraction loses the concept of message order in the agent conversation inside sets (R_n and E_n) and subsets (r_{nx} and e_{nx}) are used, but this local order is not needed for designing the agent conversation nor for the resend message. Moreover it can be restored in a detailed conception of the conversation or when the conversation is effectively introduced into the agent.

To illustrate the conversation formalism consider the following example, in the FIPA request interaction protocol (represented at the left of Fig. 5.2), the conversation corresponding is represented in the middle of Fig. 5.2 and described as follows.

For the initiator agent:

- S0, S1 and End are the states corresponding to the waiting of messages,
- $R_0 = \emptyset$ because S_0 corresponds to the beginning of the conversation by the initiator,
- $E_0 = \{\{request\}\}$
- $R_1 = \{r_{10} = \{refuse\} \cup r_{11} = \{agree; inform\text{-}done\} \cup r_{12} = \{agree; inform\text{-}result\} \cup r_{13} = \{agree; failure\}\}$,
- $E_1 = \emptyset$ because it corresponds to the last state of the conversation for the initiator,

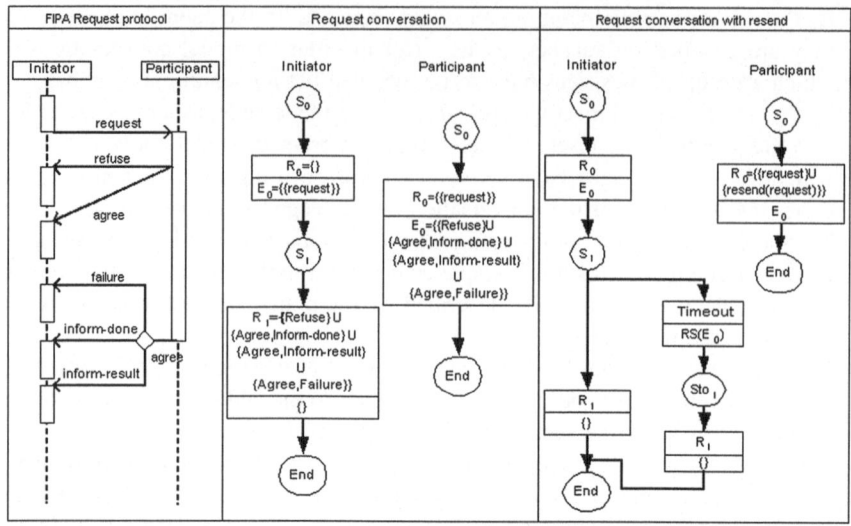

Fig. 5.2 FIPA request protocol using conversation formalism

- Leaving the state S_1 occurs when messages $\in r_{10} \cup r_{11} \cup r_{12} \cup r_{13}$ are received and no other message will be sent. The agent can receive :

  - $r_{10} = 1$ message $\in R_1$, i.e. the *refuse* message
  - $r_{11} = 2$ messages $\in R_1$ the *accept* message plus *inform-done*;
  - $r_{12} = 2$ messages $\in R_1$ the *accept* message plus *inform-result*
  - $r_{13} = 2$ messages $\in R_1$ the *accept* message plus *failure* message.

  For the participant agent:

- S_0 and End states corresponding to the waiting of messages,
- $R_0 = \{\{request\}\}$ (note that, of course, it corresponds to E_0 of initiator agent),
- $E_0 = \{e_{10} = \{refuse\} \cup e_{11} = \{agree; inform-done\} \cup e_{12} = \{agree; inform-result\} \cup e_{13} = \{agree; failure\}\}$.

Note that, with the given definition of the conversation, reception of a non-valid subset leads to a fault detection, for instance in the request protocol if the initiator agent receives the *accept* message plus *refuse* message.

Then the right of the Fig. 5.2 provides the conversation corresponding to FIPA request interaction protocol with resend, which includes the following changes into the conversation:

For the initiator agent:

- S_1 waits $R_1 \neq \emptyset$ so it has a timeout,
- when the deadline is over (timeout) the agent sends a resend(request) and goes to Sto_1,
- in Sto1 it then waits for the same R1 and has the same E1.

For the participant agent:

- $R_0 = \{\{request\} \cup \{resend(request)\}\}$.

5.4 Effectiveness

The proposed method is a time redundancy method that can therefore handle transient or temporary faults, but not in all cases. Indeed, the proposed method only succeeds if the fault is not active when the resend message is sent/received. Therefore the success of our proposed method is as strongly linked to the detection means used as for classical redundancy. For example timeout duration will have a strong impact on when the resend message is sent or received and depending on fault status (active or idle) when the resend message is sent/received the fault handling will be successful or not.

These timing issues are underlined in [12] for classical systems and or applicable for MAS. The Fig. 5.3, represents the timing issue with a sent message (e_n) arriving at time T and the associated resend message (resend(e_n) noted RS) that will be successful if arriving at T2 and not at T1. For the rest of the sections dealing with the resend handler the term temporary will be used for transient and temporary faults since the distinction is no more relevant for the issues that will be discussed.

The advantages of the resend handler compared to a retry are the following:

Resend is adapted to the knowledge level of the agents, which allows them to decide whether to use it or not as any other performative.

When an agent receives the resend performative, it provides information on the mental state of the other agent.

The internal state of the agent is not the same when receiving a resend. This avoids message repetition effects and thus corrects one disadvantage of the retry method, as well as providing the agent with the opportunity to review its internal state.

Fig. 5.3 Transient versus temporary faults

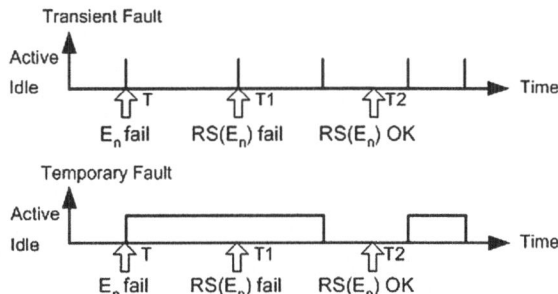

The required resend treatment will draw the designer/developer's attention to the consequences of reception of the resend message, so that he may use a preventive message about a possible delay to the requesting agent.

In the distributed system domain, for handling faults in message-passing the retry method can be supplemented by sending the conversation log (a possibly significant set of messages), which provides information to diagnose when the fault occurred and why. This is not necessary for agents, thanks to the high level of their communications (communicative acts, conversation protocols and messages structure). In particular, agents' messages are adapted to asynchronous conversations and contain information so that a specific message can be identified as belonging to a specific conversation state of a known interaction protocol. Therefore sending only the last sent message of the conversation is enough information.

Another point concerning the handling of behavioral faults is that the performative makes it possible to make the other agent change its mind. Indeed, the resend message has a specific meaning stressing the fact that the agent requires a response to an already sent message, so if the agent has chosen not to respond to the first message it can change its mind considering the message is still valid. Consider, for instance, the case of an agent involved in a Request Protocol. When it receives the request it may choose not to respond, and some time later it might receive a resend message.

The meaning of the message allows the agent to infer that:

(1) This request is not a new one;
(2) The answer is still required and thus it is important to respond to it (maybe the other agent will consider that an effort was made to respond and will return the favor).

Another noticeable advantage is that message encapsulation means that our handler cannot be confused with a mere repetition of the same message (possibly sent due to a development or malicious fault), which prevents other agents from considering the agent as malicious (when using a trust and reputation mechanism for example) or bugged.

In the case of a loss of synchronization between agents arises when messages are not received in order of transmission, or when they have been altered during transport and are thus no longer a direct expression of the expected communication protocol. A (syntactically and semantically) correct message is not received in the expected state, such fault can be detected by ignoring the message and then "wait" for a timeout or by directly going in a fault handling state by the analysis of the message, then a resend message can be sent whit the message last corresponding to the last "safe state" and finally the agents can resynchronize from this "safe state".

Finally, if the resend message is insufficient to handle the fault, an adapted handler must be chosen; but it seems that no generic handler can be provided for this situation because the solution is highly context-dependent.

5.5   Using the Resend Performative

Integrating the proposed handler into agent conversations makes them more complex, since it introduces new messages, and makes the task more tedious for the designer/developer (because it introduces new states for fault handling) as depicted in Figs. 5.4 and 5.5. But the task is not as difficult as eliminating the potential "double effect" (when a message is received twice or more) of the retry that may necessitate a diagnosis of the conversation. Sending the resend message is easy; the difficulty in using the resend performative lies in reception and treatment. Actually, a resend message is received because something in the conversation went wrong, i.e. the conversation (and therefore the agents) is no longer synchronized.

The following section presents the method we designed to incorporate the resend performative and protocol into interaction protocols.

5.5.1   Integration Resend into Agent Conversations

In the practice, the proposed protocol involves two agents: one detecting a fault and sending a message, and the other receiving a resend message. To illustrate the integration of the resend performative and protocol, the nominal conversation of Fig. 5.1 is represented with the resend handler in Fig. 5.4. The resend handler is illustrated for state S_n. States S_{n-1} and S_{n+1} illustrate the link between possible received resend messages and are not entirely represented.

The part of the resend protocol for the agent observing a fault is quite simple; the agent may use to chose the resend protocol and then must send the appropriate resend message and wait for a response. So when designing the agent conversation for each waiting state a fault detection mechanism and a next state with a handler choice must be added.

The agent may assess whether other handlers are better than the resend even if the only other known handler is to stop the conversation, because in some cases using resend may just be a waste of time. Such cases when a resend is easily evaluated as useless are, for instance, the short conversations as FIPA request, when it is the same to restart a conversation or to change the other agent involved in the conversation. In certain cases, there is no reason to use a resend; for instance in the FIPA-Contract Net protocol when the agent sends the call for proposals, in many cases it does not need all responses. If enough responses have been received the agent does not need the resend, but if the number of received responses is too low, the resend may be used.

The evaluation of the adapted handler must be done for all states but not necessary by the agent at run time, it can be done by the agent designer or developer during agent conception.

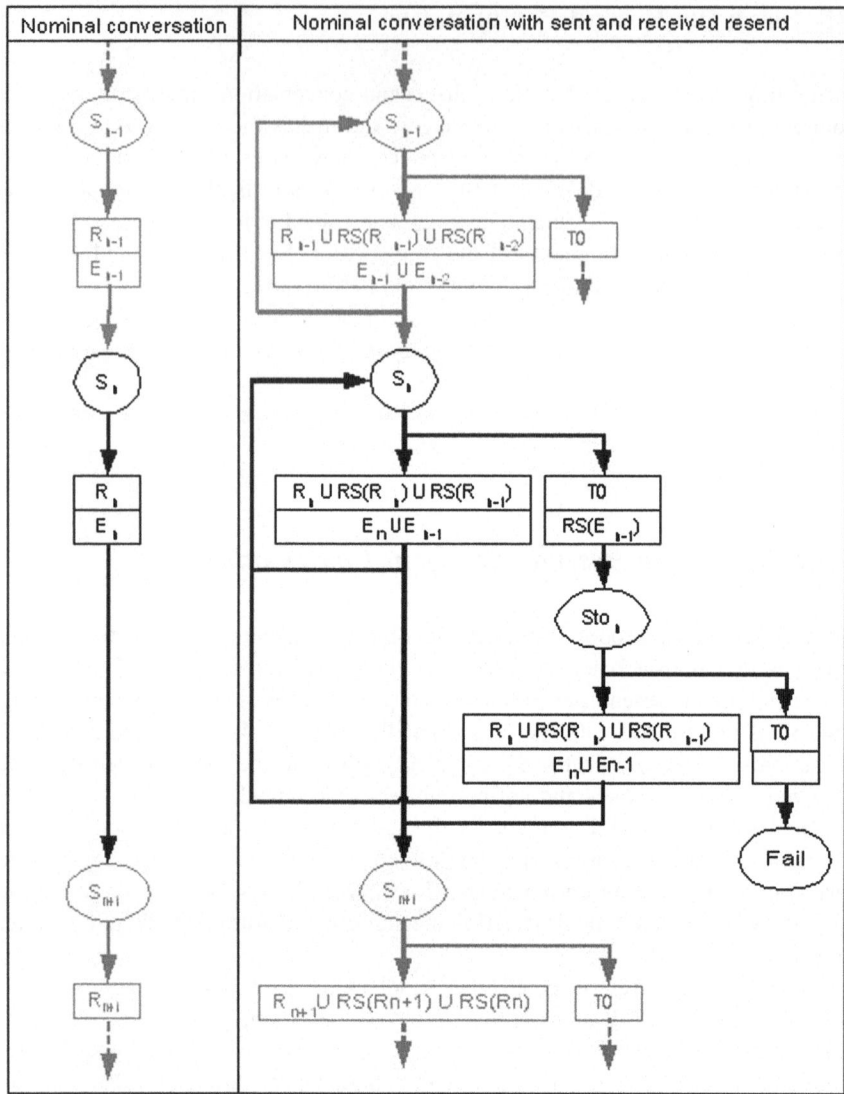

Fig. 5.4 Resend protocol on a nominal conversation

Since the resend handler implies that the agent receiving the resend can evaluate the context if an interaction protocol implies sending m messages before
receiving any (represented in the conversations by the set E_{n-1}), then all the
m messages must be sent in a resend message.

In conversation showed by Fig. 5.4 we choose to use timeout to detect faults
(all messages not into the waited E_n are ignored and then a timeout will raise if an
incorrect message is received instead of a waited one) and if a resend fail to handle

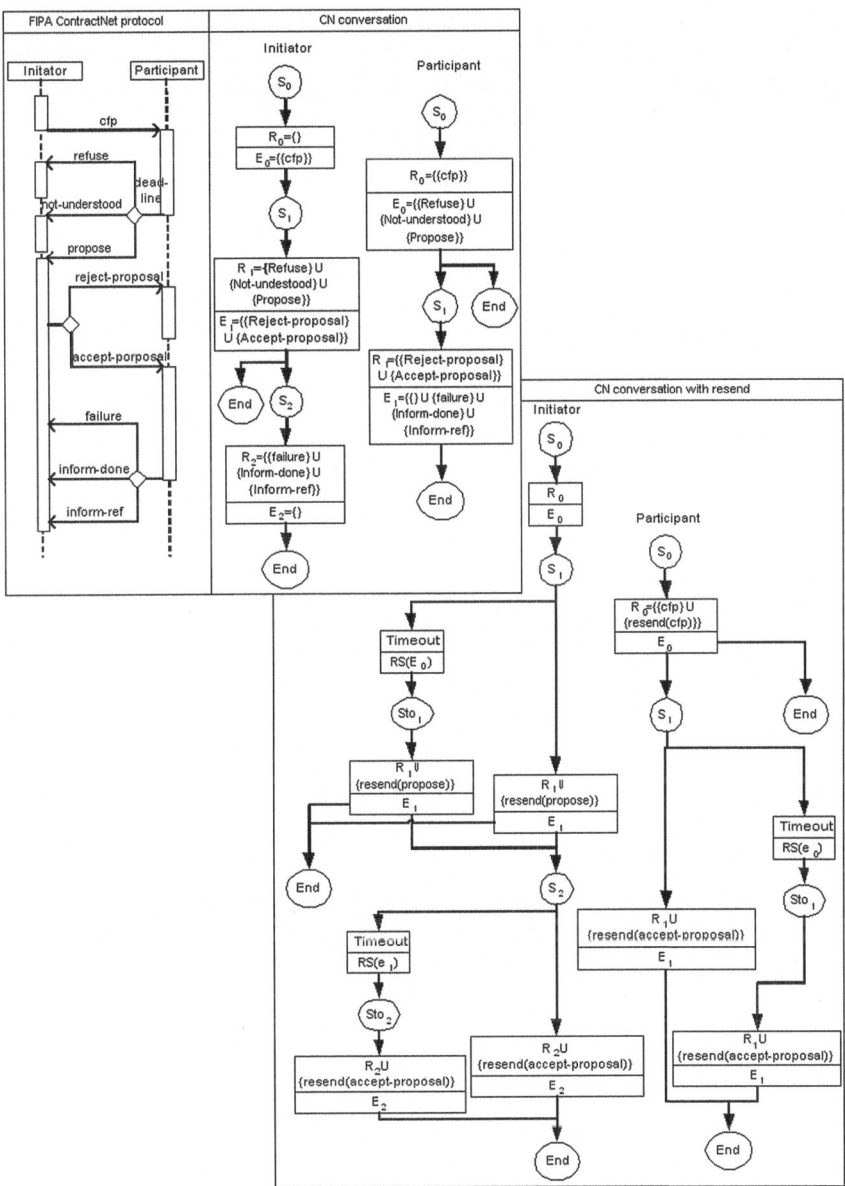

Fig. 5.5 Resend protocol on FIPA contract net protocol

the fault the conversation is ended after the second timeout. On this conversation example we notice that fault handling adds 2 new states (states "TO"), these new states correspond to "fault handling" and would be necessary for any fault handler, and resend itself does not add any state.

The other part of the resend protocol implies that the other agent is able to correctly interpret and respond to a resend message. This is the difficult part when designing agent conversation, because it means that the agent must know how to react in each state for all resend messages it can receive.

But, when using the conversation formalism, not any resend message can be received in any state, since the conversation does not go on if expected messages are not received (i.e. an agent will not move to state S_{n+1} and following, if the messages expected in S_n are not received).

For a given conversation waiting state S_n only two different resend messages can be received:

- Resend(r_n):

  - In this case, the message r_n sent by the other agent was lost, late, incorrect or ignored (i.e. agent received no message $\in R_n$).
  - An appropriate behavior can be to do as if receiving r_n instead of resend (r_n) send the appropriate e_n and go to the corresponding S_{n+1}

- Resend(r_{n-1}):

  - In this case, the response to a message r_{n-1} sent by the agent was lost, late, incorrect or ignored (i.e. the other agent did not consider a message $\in E_{n-1}$ sent by the agent).
  - An appropriate behavior can be to send the appropriate e_{n-1} since the other agent has experimented a fault after the agent sends it and go (return) to S_n.

The special state Sto_n, corresponding to the treatment of a timeout observed in state S_n, can receive exactly the same messages:

- The normal message r_n:

  - If, for example, the timeout was too short for the message to arrive.
  - An appropriate behavior can be to do as if receiving r_n in the nominal state S_n

- The resend(r_n) message:

  - If, for example, the normal message r_n was lost.
  - An appropriate behavior can be to do as if receiving r_n instead of resend (r_n) send the appropriate e_n and go to the corresponding S_{n+1}

- The resend(r_{n-1}) message:

  - If, for example, the normal message r_n was lost and the timeout was too short for the resend message of the other agent to arrive.
  - An appropriate behavior can be to send the appropriate e_{n-1} since the other agent has experimented a fault after the agent sends it and go (return) to S_n

It must be noted that r_n and e_n can represent more than one message.

For example, for the initiator agent of a FIPA request, protocol begins in state S_0 sending the message request; in state S_1, $R_1 = \{r_{10} = \{refuse\} \cup r_{11} = \{agree; inform\text{-}done\} \cup r_{12} = \{agree; inform\text{-}result\} \cup r_{13} = \{agree; failure\}\}$.

To provide an example of a more complicated interaction protocol used this resend, the FIPA contract net protocol conversation is presented Fig. 5.5 with the resend handler and the same rules explained for the nominal conversation of Fig. 5.4.

The agent receiving the resend message may evaluate it like any other message.

For example, an agent may consider that it has already treated the message r_{nx} and has no time left to do the same for resend(r_{nx}). If reception of a resend message changes the agent behavior, i.e. memory value, impact on other plans, the resend message must not be processed exactly like the message: when side effects are involved, this would result in a double effect.

The way the resend message is processed depends on when it is received, on the message context and on the need to provide a response. It is the responsibility of the agent developer to determine the consequences of reception of a specific resend message in a specific state. This is also the case for the state the agent will reach after receiving a specific resend message in a specific state.

Potentially, receiving a resend message impacts an agent entire behavior.

In particular, consider the case when a message r_{nx} had already been received in S_n, and then the associated resend message resend(r_{nx}) arrives in S_{n+1}. If the message r_{nx} has involved any changes in the agent beliefs, which is usually the case, receiving the resend(r_{nx}) message may invalidate or change this value. Moreover, time passed between reception of r_{nx} and resend(r_{nx}), so the internal state of the agent is not the same and the decision taken may change and the agent beliefs with it. Therefore the agent must determine whether the message to send in response must be the same as the message sent before, or a new message. Sending the same message avoids processing the response again (and thus saves time). It also ensures that if the first sent response reaches the other agent, it will correspond to the same information. But this message might be in contradiction with the agent's current internal state. The sending of a new assessment of the message results in a message that matches the agent current internal state but costs more time and resources. But this behavior may result in conflict if the first response sent reaches the other agent before the response to the resend. Such a problem can be addressed by defining a specific message to answer a resend, but this is not studied here.

The processing of a resend message can be the same as when the agent receives the original message, except for side effects, which must not be executed twice (messages sent, new plans begun). For large treatments, it is possible to adopt another approach, such as N-version programming that can handle some development faults that are permanent faults.

For example, if agent1 sends message r_{nx} to agent2, and agent2 has a development fault concerning treatment of r_{nx}, agent1 will receive no answer. Then agent1 uses the resend method and sends resend(r_{nx}). Agent2 will use a piece of code different from the one used for treatment of msg1, that may be a non-buggy one, and agent2 may possibly provide the expected answer.

The way the resend message is processed depends on when it is received and on the contents of the response.

First, the agent developer is responsible for finding the consequences of receiving a specific resend message in a specific state. This is also the case for the state the agent will be in after receiving a specific resend message in a specific state.

It is also the duty of the developer (or agent) to estimate whether the message to send in reply to a resend indicating that the other agent has not received a response yet, can be the same as the message sent before or a new assessment. Sending the same message avoids "calculating" the response again (saving time) and ensures that if the first response sent reaches the other agent, it will correspond to the same information, but makes it possible to send information that does not match the agent's current internal state. Sending a new assessment of the message results in a message that corresponds to the current internal state of the agent (possibly obtained through a different method, N-version programming) but takes time and may result in a conflict if the first response sent reaches the other agent.

5.5.2 Integration in the Agents

The precise method for developing agents using the proposed performative and protocol depends on the model and/or language used for agent implementation, but the chosen resend approach makes it compatible with all languages, agents architectures and exception-handling systems.

For agents using an exception-handling system or sentinels, the resend handler may be part of the known handlers.

In a MAS where agents use predefined interaction protocols, the resend handler can be implemented once for all with existing interaction protocols. This means that the final agent developer can be given a skeleton of the interaction protocol including the resend handler. To build the agent's behavior, the developer uses the interaction protocols as he wishes. He is responsible for completing the protocols with decision-making methods and other calculation operations, but need not to wonder how to use the resend method.

For ad hoc interaction protocols, the developer may have to incorporate the handler himself, using the implementation method explained before.

5.6 Discussion on Some Aspects of Resend

Some choices were made for use of the resend that were not explained previously. This section provides the reasons why we recommend sending the resend message only once, the validation of the proposed implementation method, the prospects for EMP faults not handled with the proposed method and the prospects for automating integration of the resend performative into interaction protocols.

5.6.1 Resend Only Once

Using the proposed protocol, an agent sends only one resend message and the resend message may not itself contain a resend. Why send the resend message only once per nominal message:

- Sending a message several times is a method suitable for long communication faults (unplugged network cable), as is shown in related works. These faults may be better treated by the platform itself on its own level, because they relate to message transportation and should not be the agent concern.
- If the same message is sent more than once it can be confused with a repetition possibly due to a development or malicious fault (avoided by defining the resend performative).

Why not send a resend containing a resend:

- Sending more messages to the other agent may overload it;
- The other agent is autonomous and may have a good reason for never answering or delaying its answer (other priorities);
- The other agent might be dead or have a bug, so it is useless to waste more time trying to communicate with it.

5.6.2 The Case of Event Perception Instead of a Message

The resend message is not only suitable for conversations containing only messages. It can also be used in cases where a perception is expected as the response of another agent. For example, a robot asks another robot to move because it is blocking the way. If, after a time, the first robot notices that the other has not moved it can send a resend message in the hope that the other robot will interrupt what it is doing and move.

5.6.3 Experiments Feedback

We used the resend performative in an experimental MAS designed to assess FIPA-CNP for UAV coordination. The implementation relied on a conversation-based language, the language allowed the description of agents' behavior as a set of automata, where the states correspond to waiting for a message, with each automaton dedicated to a given conversation. This served to confirm the proposed method to integrate the resend performative into existing FIPA and ad hoc interaction protocols. It also provided some hope with regard to the integration of our proposal into FT-ACL.

Moreover, the implementation was partially done by another person, which serves to "confirm" the methodology provided.

Regarding the treatment of received resend messages, our implementation confirmed the following points:

• The resend must not be integrated into an interaction protocol for all messages. Its use is strongly context-dependent.

For example, consider the interaction protocol FIPA contract net protocol; the initiator of the conversation may send a call for proposals to multiple participants, and does not generally need to obtain all responses but only a significant number. In this case, the resend performative must be used if the number of received responses is lower than needed.

• The assessment of a received resend message must take into account the side effects.
• The processing following reception of a resend message might be different from the nominal message.

For example, if the message involved changes in the agent's beliefs, reception of the associated resend message may invalidate this change. Moreover, the treatment to be carried out when receiving the resend message must not be the same as for the first message because this may imply the "multiple effect" problem underlined in the "key idea" section. This makes is impossible to automatically reuse the treatment applied to message M for message resend(M).

Experiments showed, as expected, that our protocol (if properly used) could tolerate single communication faults between agents. In fact, in real situations, the agent obtaining the expected response to a resend message depends on the probability of any fault capable of affecting the system i.e. it depends on: reliability of the communication means (loss, alteration, transmission delay), reliability of the processor (stop, overload), reliability of the agents (wrong message, wrong treatment), chances for the agent to change its decision (autonomy, which may depend on many criteria unknown to the agent sending the resend message).

The next section presents possible applications of our classification and factorization of faults.

References

1. Avizienis, A., Laprie, J.-C., Randell, B., Landwehr, C.: Basic concepts and taxonomy of dependable and secure computing. In: I. computer society, editor, IEEE Transactions on dependable and secure computing, pp. 11–33. (2004)
2. Cristian, F., Fetzer, C.: The timed asynchronous distributed system model. In: FTCS, pp. 140–149. (1998)
3. Klein, M., Dellarocas, C.: Exception handling in agent systems. In: Etzioni, O., Müller, J.P., Bradshaw, J.M. (eds.) Proceedings of the Third International Conference on Autonomous Agents (Agents'99), pp. 62–68. ACM Press, Seattle (1999)

4. Barbuceanu, M., Fox, M.S.: Cool: A language for describing coordination in multiagent systems. In: Lesser, V., Gasser, L. (eds.) Proceedings of the First International Conference oil Multi-Agent Systems, pp. 17–24. AAAI Press, San Francisco (1995)
5. Koren, I., Koren, Z., Su, S.Y.: Analysis of a class of recovery procedures. IEEE Trans. Comput. **35**(8), 703–712 (1986)
6. Dastani, M.: 2APL: a practical agent programming language. Int. J. Auton. Agents Multi-Agent Syst. (JAAMAS). **16**(3), 214–248 (2008) (Special Issue on Computational Logic-based Agents, (eds.) Francesca Toni and Jamal Bentahar)
7. Dragoni, N., Gaspari, M.: Crash failure detection in asynchronous agent communication languages. Auton. Agent. Multi-Agent Syst. **13**(3), 355–390 (2006)
8. Potiron, K., Taillibert, P., Fallah-Seghrouchni, A.E.: A new performative for handling lack of answers of autonomous agents. In: ICAART, pp. 441–446. (2009)
9. Searle, J.R.: Speech Acts: An Essay in the Philosophy of Language. Cambridge University Press, Cambridge (1969)
10. D. T. FIPA. FIPA communicative act library specification, 2001
11. Finin, T., Labrou, Y., Mayfield, J.: KQML as an agent communication language. In: Bradshaw, J. (ed.) Software Agents, MIT Press, Cambridge (1997)
12. Guessoum, Z., Faci, N., Briot, J.-P.: Adaptive replication of large-scale multiagent systems: towards a fault-tolerant multi-agent platform. In: Proceedings of the fourth international workshop on Software engineering for large-scale multi-agent systems, pp. 1–6. ACM Press, St. Louis, Missouri, (2005)
13. Potiron, K., Taillibert, P., Fallah-Seghrouchni, A.E.: Autonomous agents: When the mailbox remains empty. In: IAT, 2009

Chapter 6
Fault Classification Attributes as an Ontology to Build Fault Tolerant MAS

Abstract Some handling methods had been studied into Multi-Agent System (MAS) domain, adapting to their specificities and capabilities but increasing the large amount of fault tolerance methods. Therefore, unless being an expert in fault tolerance, it is difficult to choose, evaluate or compare fault tolerance methods. This prevents a lot of developed applications not only to be more user friendly but more importantly to, at least, be tolerant to common faults. As stated in the introduction, our goal in building this fault classification was to find a way to design fault-tolerant MAS. The fault classification presented in this book identifies the faults MAS may be subject to, and also permits MAS designers and developers to specify fault tolerance with regard to the required properties. Moreover this book also submits the possibility of unifying the terms used by MAS researchers in describing what their work addresses.

Keywords Multi-agent system · Dependability · Fault tolerance · Multi-agent system design · Behavioral faults

6.1 Fault Specification Phase

The need to integrate dependability processes into the development model was, for instance, presented in [1] and will therefore not be argued any further. A first step into this, and a significant piece of work, for a MAS designer is to analyze which faults the system (i.e. the agents, the platform and maybe also the hardware) must be tolerant to. For a MAS designer, the first step to do so—a significant piece of work—is to analyze which faults the system (i.e. the agents, the platform and possibly the hardware) must tolerate. A method to complete this task is first to specify which faults the system may experience during its lifetime and then to dispatch handling of the faults between agents and platform, before choosing which handlers may be more convenient for each potential fault.

K. Potiron et al., *From Fault Classification to Fault Tolerance for Multi-Agent Systems*,
SpringerBriefs in Computer Science, DOI: 10.1007/978-1-4471-5046-6_6,

The difficulty of specifying exactly which faults the system may experience, which consequence every simple and multiple fault may have on other parts of the system, and how to detect such faults and then respond to them, is directly proportional to system complexity.

We will examine here the possibility of working on fault classes instead of every individual fault. In [4] the author says that the first prerequisite when designing fault tolerance is to specify the faults to be tolerated, which is done more concisely by specifying fault classes than faults. But he provides no fault classes and no explanation on how to do so. Our argument is that using 37 fault classes to represent all possible faults requires understanding the attributes, their values and what they represent, but this is worthwhile compared with enumerating every single fault the system may experience.

We, then, give guidelines and examples to determine the MAS system specification using fault classes. This specification relies on the use of the fault classification and can be done in two stages: first, identifying the fault classes the entire system may be subject to, and then, which fault classes must be addressed by the platform and which ones by the agents, some being possibly addressed by both of them. This is the organization of the upcoming sections.

6.1.1 The Specification Phase of the Entire System

The set of faults to be tolerated by a system depends on its usage, but also on the system considered. Here, the term "system" is used in its widest sense: the hardware and software are under consideration because 37 fault classes entirely cover such systems.

Here, we will be considering the system defined below, see also Fig. 6.1:

- **System point of view**: the MAS (agents and platform, discussed in the next section) and the required hardware, i.e. "the entire thing being built";
- **Environment**: the part of the "outside/external world" that can impact or be impacted by the system (the definition can be more precise when studying a real device);
- **Boundary**: system communications and/or hardware interfaces with the environment including users;
- **User(s)**: part of the environment that uses the system: other agents, MAS and/or human being(s).

If development faults can hardly be ignored, physical and interaction faults are not always relevant for applications. Moreover the amount of work to tolerate physical and interaction faults is not always worth valuing their impact on the system. For example, when designing a system to display blogs on Internet, tolerating interaction faults includes guaranteeing that each user request for information reaches the server.

Fig. 6.1 System considered at the beginning of the analysis

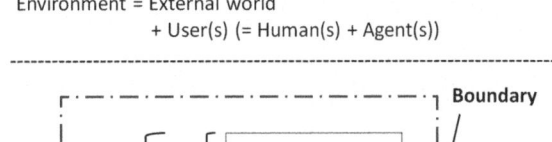

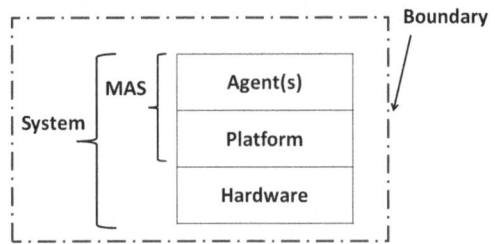

It is too expensive compared to the small inconvenience for the user, who can repeat the request. Transmission problems are too costly to guarantee on the Internet, so they will not be considered. Whereas for selling books, because of the payment procedure a transmission fault costs more than using fault tolerance and so will be taken into consideration.

Which fault classes have to be taken into account will depend on the kind of system (open, closed), the user's fault tolerance requirements, the environment where the MAS will be executed, the kind of components, hardware requirements, development costs, etc. Note that a wrong decision in these specifications results in the creation of some development faults (in fault classes numbers 1–3 and 6–8).

Since it is impossible to list exhaustively the criteria that might be considered in order to ignore some fault classes, we present, below, the most exhaustive list we could make for fault classes that can be handled optionally and the corresponding circumstances:

- If the system considered does not include hardware,

  - Then faults 5–9 plus 10 and 11 are off the subject because they are internal hardware faults.

- If the system considered includes hardware but is not subject to a difficult environment (for example, a schedule-management software),

  - Then the corresponding physical fault classes 12 and 13 can be ignored; for instance this will not be the case for a system supervising a production line.

- If the system considered includes hardware but no human can harm it (e.g. access to hardware limited to trusted humans),

  - Then fault classes 14–19 can be ignored.

- If agents cannot cause physical faults then fault classes,

- Then 26, 27, 36 and 37 have no reason to be taken into consideration by the agent designer.
- If the system runs on a single computer,
  - Then faults related to message transmission can be withdrawn (faults 12 and 13), this will not be the case for Internet applications.
- If the system has no interactions with human beings (a MAS acting as a server may interact only with other client MAS (humans use another MAS for direct interaction) or a MAS interacting with dedicated software), or if the human users have no interest in cheating,
  - Then the fault classes 20–25 can be ignored.
- If system developer(s) are fully trusted,
  - Then the development fault number 4 can be ignored.

Such a selection among the fault classes that the system will experience makes it possible to focus on identifying only which faults in the remaining classes require a deeper analysis. But the possible use of fault classes is not limited to the specification at the system level: it can also be used to dispatch fault treatment among system components.

Indeed, the system considered is composed by entities that will certainly not have to, or be able to, tolerate all selected fault classes. So we will next discuss how the fault classes may be dispatched to the corresponding entities.

6.1.2 The Dispatching Phase Among the System Entities

After defining the fault classes the system must tolerate, we must determine which "component" may be affected by which faults, or which "component" can more efficiently handle which faults. This is what we name the dispatching phase.

We will not discuss which faults can be handled by the hardware since we consider MAS here as a software only system. Moreover, we would only be able to say that potentially all faults can be addressed by special/adapted hardware items.

When considering the software part (MAS part on Fig. 6.1.) of the whole system under investigation (defined on Fig. 6.1.), one step involves defining which of the fault classes, among those isolated before, must be handled by the platform.

A platform here is defined as all the pieces of software needed to allow agents to live their lives, e.g. the operating system, means of direct communication between agents, indirect communications between agents (e.g. pheromones), agent managers, directory facilitator, etc.

To define which faults concern the platform we must redefine the system under consideration. It changes, for this study, the role of the platform, agents, hardware and all components of our whole system, as shown below and in Fig. 6.2:

- **System/point of view**: MAS platform;
- **Environment**: agents and hardware;
- **Boundary**: system interfaces (with agents and with hardware);
- **User(s)**: agent(s), human(s).

When taking into account all the 37 possible fault classes, we can consider the platform responsible for handling:

- Development faults, represented by faults 1–9;
- All software development faults since the platform is a software program;
- All hardware development faults because of the platform, since logic-wise, the low-level program is well suited to have information on hardware;
- All Hardware operational faults, represented by faults 10–19 plus 26, 27, 36 and 37, since the platform can have some "low-level" information and some control on hardware. For example, an overloaded network or unplugged network cable might be easier to detect, diagnose and handle at platform level, since agents are not be responsible for message transmission.

Then, we will be considering the other software entities of the system (agents), see Fig. 6.3:

- **System/point of view**: one or more agent(s);
- **Environment**: platform and other agents and hardware;
- **Boundary**: agents boundary;
- **User**: agents (maybe from another MAS), human user.

Let us note that one agent or one kind of agent may need a separate study, for example when they have a specific role as agents regulating the system interface with the user.

In such a case, one kind of agent may be responsible for handling several fault classes that another agent will not handle at all.

We can consider that the agents (all of those in the considered MAS) will have to handle the following faults:

Fig. 6.2 MAS platform under consideration

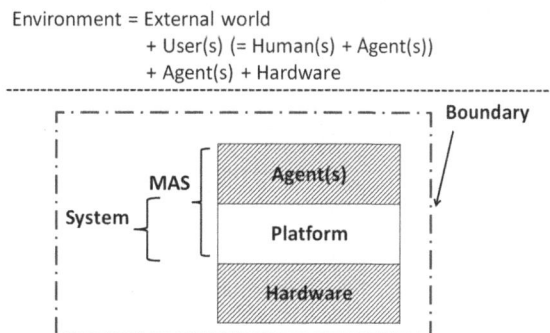

Environment = External world
+ User(s) (= Human(s) + Agent(s))
+ Agent(s) + Hardware

Fig. 6.3 Agent(s) under consideration

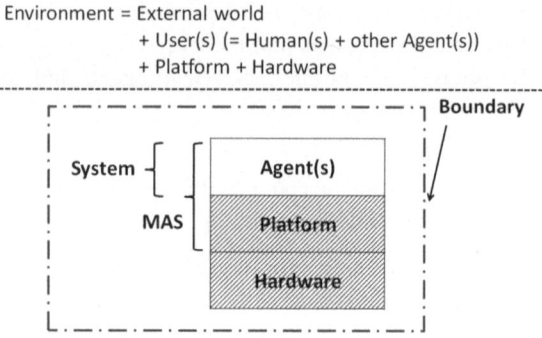

Environment = External world
+ User(s) (= Human(s) + other Agent(s))
+ Platform + Hardware

- Development faults represented by faults 1–4, since the agents are software items;
- Interaction faults except the hardware ones, these cases (faults 20–25) containing for example all interactions with human beings or other programs that may be handled by the agent at its level;
- Behavioral faults corresponding to the faults 28–35, mostly when considering one agent interacting with other agents.

The analysis of the faults to be handled by the platform and the agents is summarized in Table 6.1 for the analysis we use only the values of the two most discriminative attributes (*phase of creation or occurrence* and *dimension*) when all faults are taken into consideration.

For example, a fault being deliberate or not is not a discriminative attribute when making such an analysis.

Only the fault classes 1–4 have to be handled by the platform and the agents since they are development software faults.

This book only gives a high-level analysis because it is impossible to detail all the combinations of all platform or agent fault tolerance requirements. It is based only on two classification attributes (*phase of creation or occurrence* and *dimension*) since they appear to be the top two discriminating ones for platform and agents. To refine the analysis, other attributes will be required: for instance *intention* or *phenomenological cause* which appear to be the next two more discriminating attributes.

Table 6.1 Corresponding faults for the platform and the agents

	Faults						
	Development		*Operational*		*Autonomy*		
	Software 1–4	*Hardware* 5–9	*Hardware* 10–19	*Software* 20–25	*Hardware* 26 and 27	*Software* 28–35	*Hardware* 36 and 37
Platform	x	x	x		x		x
Agent	x			x		x	

6.1.3 Comments on the Specifications

When using a closed platform (without the ability to change anything in it), the issue is simple: the platform will handle only the faults its designer has picked. This can be an important criterion in the choice of a platform.

In this case, all remaining faults from MAS specifications must be handled by agents, which may result in:

- Problems handling certain faults because the platform does not provide agents with the means or information to handle them;
- A need to review the faults that must be taken into consideration for the system as a whole because some faults cannot be handled at agent level.

A platform that does not handle the required faults can result in a MAS that cannot handle all required fault classes or only at a very high price in terms of program complexity (an important source of development faults).

Realizing that the choice of platform is linked to the faults it can handle, platform designers should provide an analysis of the faults their platform handles, even if the platform was not designed with fault tolerance in mind. Without such an analysis, the industrial use of MAS will always require creating a new platform that will fulfill its needs as regards dependability.

Moreover, if the platform is associated with a strict agent model (agent designers may only change an agent behavior or the agent capabilities are limited/restricted), the faults can be handled by adapting the behavior. This may result in a more complex behavior for the agent (another source of development faults) and a true limitation to the type of faults that can be handled if the agent model provides nothing as regards fault tolerance. Such agent models will be less appreciated, especially for critical systems, than those providing real tools to deal with fault tolerance issues.

Note that the use of a temporal model, such as the timed asynchronous model [3] or the synchronization of asynchronous entities [9], implies specific hypotheses concerning the faults.

For example, the following assumptions are made for fully asynchronous systems: services are time-free (an output always occurs in response to an input without any boundary in time); inter-process communication is reliable (a condition sometimes relaxed); processes only have crash failures and have no access to hardware clocks.

Some of these assumptions imply the need for strong guarantees, particularly regarding message transmission and process failures. They will require using appropriated fault tolerance for message transmission and process failure, but much less is needed in the process behavior to handle external faults.

In another example of a temporal model, the timed asynchronous model, the assumptions are: all services are timed; messages can suffer omission or performance failures; processes can crash or have performance failures; processes must have access to hardware clocks and no bounds are set on communication frequency and process failures.

These assumptions imply fewer guarantees on required fault tolerance for message transmission and process failures, but will need more tolerance in the process behavior to handle external faults.

Let us note that even for software-based systems on automatically-generated code including some fault-processing, the system designer must still be interested in doing the analysis to define, in the specification, which faults remain to be taken into consideration.

The layered approach that comes with MAS considering the platform responsible for "low-level faults" and agents responsible for "high-level faults", appears to be an argument for MAS fault tolerance, because when the platform is well suited to the requested fault tolerance level, the agent designer/developer has to work less to create fault tolerant agents.

After specifying the faults that must be handled at each level of the MAS, the designer must know how to choose the appropriate fault handlers to incorporate them in the system design.

6.2 Handler Specification Phase

The term handler is used here to represent any method permitting to handle faults.
Obviously, different faults can lead to the same errors, for instance:

- Agents can possibly not respond;
- The network can lose messages or
- Agents are vulnerable to development faults resulting on wrong interaction protocol.

These faults are equivalent as regards the resulting error: no message is received. Since the fault classification is useful to identify the system and agents fault tolerance needs, it may be interesting to study which faults can be handled by the same handler. Note that since the fault classification is concerned with every kind of fault, the handlers chosen will be conceptual methods or programming handlers (exception handling, for instance).

To illustrate the handlers analysis that can be done to determine which faults they can handle, we will first give a comparison between time redundancy used when exchanging messages and a performative we defined for agents to express an acknowledge of no-reception. Then we will provide the results of other handler analysis.

6.2.1 Handlers Comparison: An Example

We classified resend and time redundancy with regard to the values of the attributes of the faults they can best handle. Note that the analysis below and the table present the values for which the handlers are potentially suitable. We do not guarantee that they can handle every single fault with the corresponding values. The results are summarized in Table 6.2.

The behaviors of agents using time redundancy and then resend handler is presented before each analysis to sustain the understanding with a simplified version of conversation formalism. In these figures a simplified version of the conversation formalism defined Sect. 5.3.4 is used:

- "S(n)" stands for "waiting State n"
- "M" stands for "received Message"
- "RepM" stands for "sent response"
- "Timeout" stands for "detection mean for no response"

The *time redundancy*, see Fig. 6.4 representing the behavior of an agent using time redundancy, is possibly suitable for the following values:

- **Phase of creation or occurrence**:
  - suitable for *operational* faults such as a loss of a message,
  - not suitable for *development* faults since the same message will be treated in the same way,
  - not suitable for *autonomous* behavior since sending the same message will not make the agent change its internal state as it provides no new information).

- **System boundaries**:
  - suitable for *external* faults since the method is designed to handle faults at the communication level.

- **Dimension**:
  - not suitable for *software* faults for the same reason it is unsuitable for *development* faults.

- **Phenomenological cause**:
  - not a discriminating attribute, since the method is suitable for all values.

- **Objective**:
  - suitable for *non-malicious* faults, since any cooperative agent will try to help with the fault.

- **Capability**:
  - not suitable for *deliberate* faults since it does not imply any possible change in the internal state of the other agent.
  - not suitable for *incompetence* faults since it will always be treated through the same procedure.

- **Persistence**:
  - not suitable for *permanent* faults since sending the same message is useless when nothing changes in the environment.

Table 6.2 Summary table for handled faults of *retry* and *resend*

	Phase of creation or occurrence		Autonomy	System boundaries		Dimension		Phenomenological cause		Objective		Capability			Persistence	
	Development	Operational	Autonomy	External	Internal	Software	Hardware	Natural	Human-made	Non Malicious	Malicious	Accidental	Deliberate	Incompetence	Permanent	Transient
Retry		x	x	x			x	x	x	x		x				x
Resend	x	x	x	x	x	x	x	x	x	x			x			x

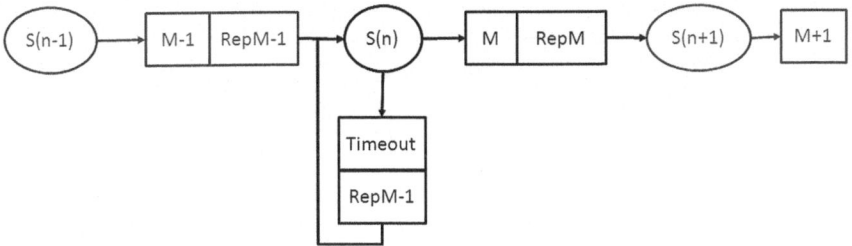

Fig. 6.4 Overview of *retry* in agent behavior

The *resend handler*, see Fig 6.5 representing the behavior of an agent using the resend handler, is possibly suitable for the following values:

- **Phase of creation or occurrence**:
  - suitable for *operational* faults such as a loss of a message,
  - suitable for *development* fault since the message will be treated as a new message and therefore through another procedure,
  - suitable for *autonomy*, since sending a message corresponding to an expressive speech act [10] have a different meaning than sending the same request, and will influence the internal state of the agent because it provides new information.

- **System boundaries**:
  - suitable for *external* faults since the method is made to handle faults at the communication level.

- **Dimension and Phenomenological cause**:
  - not discriminating attributes since the method is suitable for all values of the attributes.

- **Objective**:
  - suitable for *non-malicious* faults, since any cooperative agent will try to help with the fault.

- **Capability**:
  - suitable for *accidental* faults since some of these faults are temporary,
  - suitable for *deliberate* faults since it implies a change in the internal state of the other agent,
  - not suitable for *incompetence* faults since it will always be treated with the same abilities.

- **Persistence**:
  - not suitable for *permanent* faults since sending the same message is useless when nothing changes in the environment.

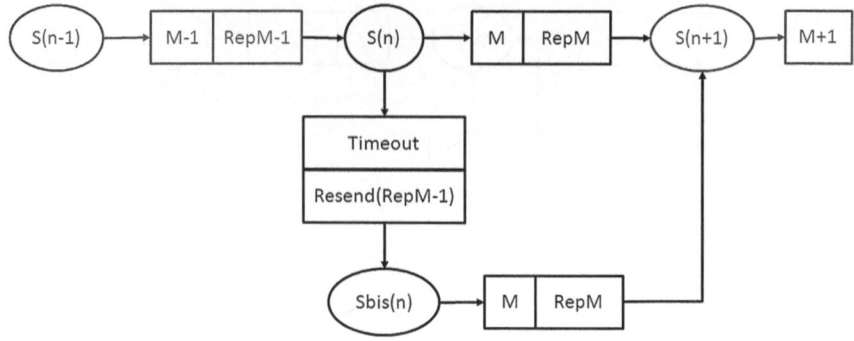

Fig. 6.5 Overview of *resend* in agent behavior

6.2.2 Analysis of Handlers and Detection Means

We were interested in studying a set of non-specialized handlers to emphasize the faults they are adapted to deal with. We first tried to assign to every handler the fault classes it can deal with, but this was not satisfying because it did not show why a fault class would be treated by one set of handlers and by not the others.

This is why, as for the handler comparison, we chose to use the values of the fault classification attributes to analyze the handlers. These values become an ontology to describe handler characteristics with regard to the faults they can deal with.

Afterwards, we made a presentation of the handlers we chose to consider in Tables 6.3 and 6.4 the first table representing methods strongly related to agents and the second table representing methods used for conventional systems. The results of the study are presented in Table 6.5.

Table 6.5 shows that, contrary to what could be expected, fault tolerance methods used for conventional systems cannot guarantee the handling of behavioral faults (only one of them, N-version programming, allows tolerance of faults linked to autonomy).

Table 6.3 Agents handlers description

Name	Description
Resend	A method designed for agents to handle some interaction faults based on the argument that a retry method can be used in a cooperative way if adapted to a high level(presented in [8]).
Acquaintance change	A possibility for the agent to replace the agent with which it is making a transaction to obtain the result it needs.
Trust	Methods intending to evaluate if the interacting agents can be trusted or not.
Re-planning	A method used to adapt the internal plans of the agents to unexpected situations.

Table 6.4 Conventional handlers description

Name	Description
Time redundancy	A method, meaning trying the same program instruction since the expected result is obtained [6].
Hardware redundancy	A method, consisting on replicating on different hardware the considered agents, so that if the hardware or link fails another copy of the agent on another hardware can replace it [7].
N-version	A method used in classical fault tolerance, the system runs different implementations of the same considered software potentially on the same hardware [7].
Rollforward	A method based on the research of a new safe state in all the possible states of the agent. For example it is possible to restart the agent and to try to have new pieces of information on its environment [2].
Rollback	A method consisting of saving from time to time a safe state of the agent so that in case of failure the agent can be restarted from the last safe state [2].

This can be explained by the fundamental differences between assumptions made when interacting with autonomous agents and assumptions made when considering conventional systems. It points out the fact that agents need handlers adapted to their specific faults, but that the handler definition must be as generic as possible to reduce the difficulty in handling behavioral faults.

When designing a system, more than one handler can manage a specific class of fault, therefore the cost (resources required) and difficulty/ease in integrating the handler will help determine if one or more of these handlers will be used.

After classifying handlers using the fault classification ontology, we will study it as a means to choose, during system execution, the handler to use.

6.2.3 Choice and Monitoring of Handlers

This section is a synopsis of the final use of the classification we have in mind in order to choose and monitor the handlers.

After a fault diagnosis, an agent (or sentinel) would have to choose some handlers to manage the faults. This could be another way of using this classification. It is possible to classify handlers with the same attributes and values as the faults they handle. In doing so, the choice of the handler can be made by matching a diagnosed fault's properties to the classification of handlers during runtime. And using such an ontology also allows exchanging handlers between agents. But diagnosis may be quite difficult for agents because of other agent autonomy so the question becomes: is it possible to diagnose the behavioral faults of other agents?

Like other programs, MAS would have to make some assumptions about the faults they face, because of the difficulty of diagnosing the faults precisely. After doing so, they will be able to choose the corresponding fault handler. For practical purposes, general diagnosis methods exist. This point is worth studying.

Table 6.5 Handlers analysis

	Phase of creation or occurrence			System boundaries		Dimension		Phenomenological cause		Objective		Capability			Persistence	
	Development	Operational	Autonomy	External	Internal	Software	Hardware	Natural	Human-made	Non Malicious	Malicious	Accidental	Deliberate	Incompetence	Permanent	Transient
Resend	x	x	x	x		x	x	x	x	x		x	x			x
Acq.change		x	x	x		x	x	x		x	x	x	x	x	x	
Trust		x	x	x		x		x	x		x		x		x	x
Replan	x	x	x	x	x	x		x	x		x		x	x	x	
Time redund.		x		x			x	x	x	x		x	x			x
Hardware redund.		x			x		x	x	x	x		x	x	x	x	x
N-version	x	x		x	x	x		x	x	x	x	x	x	x	x	x
Rollforward	x	x			x	x		x	x	x		x				x
Rollback		x			x	x		x	x	x		x	x			x

Diagnoses have two main uses: fault removal and behavior adaptation. In the case of behavioral faults, the first has no meaning, but diagnosing these faults can be quite useful for behavior adaptation.

Diagnosing the agents specific faults is a true problem since they are autonomous, and it is not stated whether sentinels [5] can have such a model of agent behavior that they can diagnose/recognize an act of autonomous behavior from a development fault (in Sect. 4.5 we showed that these faults are quite similar).

Moreover, not all MAS use sentinels, and doing so, the agents must try to make the diagnosis "by themselves" (internally). Because they have only a partial view of their environment, it is quite difficult for them to obtain some of the necessary diagnosis information, and much more difficult when interacting with non-cooperative agents.

How can the system distinguish a faulty agent from an adaptive agent, since in both cases the agent's behavior is not entirely foreseeable?

Even if the platform has to provide them with information on low-level faults when the agents need it (issues with communication links, etc.), with this information the agents can only diagnose physical faults, and how whether the experienced fault is physical or not. Otherwise it can try to make a diagnosis in the remaining subset of faults. However, what makes it possible to find a development fault or to explain that an agent uses its autonomy issues?

We do not have an answer yet. But these reflections lead us to consider that maybe MAS will need to step away from conventional system dependability. MAS may need to address dependability in a new way that may bring conventional system dependability the new lease of life it needs, because it is becoming increasingly difficult to certify dependability requirements.

References

1. Avizienis, A.: Toward systematic design of fault-tolerant systems. Computer **30**(4), 51–58 (1997)
2. Avizienis, A., Laprie, J.-C., Randell, B., Landwehr, C.: Basic concepts and taxonomy of dependable and secure computing. IEEE Trans. Dependable Secure Comput. **1**(1), 11–33 (2004). (IEEE computer society, editor)
3. Cristian, F., Fetzer, C.: The timed asynchronous distributed system model. In: FTCS, pp. 140–149 (1998)
4. Gartner, F.C.: Fundamentals of fault-tolerant distributed computing in asynchronous environments. ACM Comput. Surv. **32**(1), 1–26 (1999)
5. Klein, M., Dellarocas, C.: Exception handling in agent systems. In: Etzioni, O., Müller, J.P., Bradshaw, J.M. (eds.) Proceedings of the Third International Conference on Autonomous Agents (Agents'99), pp. 62–68. ACM Press, Washington (1999)
6. Koren, I., Koren, Z., Su, S.Y.: Analysis of a class of recovery procedures. IEEE Trans. Comput. **35**(8), 703–712 (1986)
7. Laprie, J.-C., Béounes, C., Kanoun, K.: Definition and analysis of hardware and software-fault-tolerant architectures. Computer **23**(7), 39–51 (1990)
8. Potiron, K., Taillibert, P., Fallah-Seghrouchni, A.E.: A step towards fault tolerance for multi-agent systems. In: Languages, Methodologies and Development Tools for Multi-agent

Systems First International Workshop, Revised Selected Books Lecture Notes in Computer Science, vol. 5118, 4–6 Sept (2007)

9. Schneider, F.B.: Synchronization in distributed programs. ACM Trans. Program. Lang. Syst. **4**(2), 125–148 (1982)

10. Searle, J.R.: Speech Acts: An Essay in the Philosophy of Language. Cambridge University Press, Cambridge (1969)

Chapter 7
Conclusion

Abstract This section concludes the book giving a summary of the propositions detailed all along the previous sections and then an outline of some possible future works.

Keywords Multi-agent system · Dependability · Fault classification · Autonomy · Behavioral faults · Resend

7.1 Summary

This book has shown that autonomy and pro-activeness raise the need for a significant extension to conventional systems fault classification. To do so, it drew the consequences from the fact that agent behavior is not entirely foreseeable for other agents. This implies that agents making their own decisions can, voluntarily or not, be responsible for faults that other agents will experience.

This book then pointed out the relevant faults and demonstrated which specific faults are possible for Multi-Agent System (MAS). Autonomy was added as a value of the *"phase of creation"* attribute, to represent faults resulting from autonomous behavior. A new group of faults, named behavioral faults, was defined and the 12 faults it contains have been analyzed in depth.

Then a specific new handler designed specifically for agents has been presented.

Finally, this book pointed out some potential uses of our classification as a tool for designers, to determine which faults the agents must handle in order to interact with autonomous agents in a dependable way and presented a complete study and comparison of two generic handlers.

K. Potiron et al., *From Fault Classification to Fault Tolerance for Multi-Agent Systems*,   75
SpringerBriefs in Computer Science, DOI: 10.1007/978-1-4471-5046-6_7,
© The Author(s) 2013

7.2 Future Work

Further work must be done to provide a complete analysis of the faults that may be handled by the agent or the platform, to study an exhaustive list of generic handlers and to find other tools needed to obtain fault tolerance as a property of MAS with as little effort as possible from the designer. As this study pointed out the fact that the help designers and developers received from the fault classification is insufficient to resolve the issues also encountered in conventional systems. Future work will also have to focus on the choice and monitoring of handlers during system runtime, which is the last use of our classification we foresee. It is possible to classify handlers with the same attributes and values as the faults they handle. Then the choice of the handler can be made at runtime by matching diagnosed fault properties to the handler classification.

Moreover, using such ontology to describe handlers and diagnose faults also allows exchanges between agents.

7.2.1 Unforeseen Faults

An unforeseen fault is a fault that was not included in the specifications of the system at the time of the system design and thus will not be addressed by a specific handler. This is the case of faults which were not identified at design or whose handling was deemed unnecessary (e.g. because they were estimated to have a low probability of occurrence or because the costs of handling them were considered unacceptable with rapport to the benefits). As the complexity of applications increases, even with safety assessment the chances of overseeing fault chains and the costs of tolerating such faults increase dramatically. And as fault tolerance is usually achieved through specific behaviors for all identified faults and fault chains, the ones that "get away" can lead to unforeseen faults that occur for systems that are already in operation and so to catastrophic disasters.

The aim of the research of Costin Caval is to create systems that can tolerate unforeseen faults in the sense that they detect abnormal behaviors and attempt to use the unaffected resources to continue to function as close as possible to its specifications. This research focuses on the errors resulting from the manifestation of such faults and the corrective measures to be taken to ensure the best response in the given circumstances by reconfiguring using the available redundancies.

The proposed approach reunites the four following steps:

1. Error detection, based on constraints and design by contract to help identify behavioral deviations corresponding to unforeseen faults.
2. Error confinement, by stopping the activities of an agent that detecting a problem that affects its functioning.
3. Localization of elements that were affected by the error by keeping track of the collaborations within agents. If later one of these agents is identified as faulty

the consequences of its actions will be traceable (e.g. all agents that used corrupted data from an initial faulty agent can be identified).

4. System reconfiguration according to its specific goals avoiding elements identified as faulty or "contaminated". Depending on the situation, a reparation step can be considered for regaining previously blocked resources (e.g. restarting agents in order to discard corrupt data)

7.2.2 Diagnosis for the Choice of a Fault Handler

To make a decision on whether to use resend or another handler, the agent needs some information about the fault. Since resend is not adapted to permanent faults it may be useful for an agent to detect them so that no time is lost trying a resend when it is ineffective.

Moreover, even if the resend handler succeeds in reestablishing the agents' nominal conversation, they may usefully share the information they each have about the fault, for example in case of a too-short timeout.

Using the log of timeouts and some rules, we were able to diagnose simple fault cases at little cost to the agents (and no cost to agent developers). Some examples of diagnosed faults:

- Development fault in the interaction protocol (incorrect instantiation of the protocol or fault in the decision) of an acquaintance.
- Internal development fault concerning a protocol.

Such diagnoses are quite simple but it is possible to provide adapted handlers, which limits the work to be done with regard to fault tolerance when designing the agent.

Adding to those diagnoses, the log of resend messages may provide more information with regard to the moment when the resend message is received.

- Receipt of a Resend during a "normal state": a message sent to the other agent was lost or ignored by itself.
- Receipt of a Resend during a "resend state": the sent response was lost or ignored, possible loss of two messages.

These diagnostic possibilities were not studied anymore because of a lack of time but using the adapted handler is an important issue since a wrong handler implies at best a loss of time and at worst some new faults.

7.2.3 Resend Automation

Since the resend integration is a tedious task, one can consider automating its treatment. This does not seem to be an easy task because of the choices that have to be made for the integration into the conversations and the impact of resend messages on agent behavior.

Using the proposed conversation model automating the translation into automata seems to be possible, but we have not done it for the moment. If the automata can be generated automatically providing conversations skeletons, use of hooks looks a promising way for developers to provide the agent behavior (all decision making and calculus).

For example, consider an agent which, after receiving a message M, starts a new plan while continuing the conversation. When it receives a message resend(M), this may impact the other plan but if the resend handler had been automatically included in the conversation protocol, this fact is hidden from the developer.

Those possibilities were not studied at all but may provide a great help to multi-agent system designers and developer facilitating the focusing on complex fault chains.

7.2.4 Prospects About Persistent Faults

We aimed to address the general case for EMP. This chapter focuses on temporary faults and provides a first approach towards generalization. It must be noted that if the resend protocol fails (persistent faults), then other alternatives are still possible, including the following:

- Change its acquaintance, i.e. replace the agent with which it is making a transaction to obtain the result it needs.
- Replan, i.e. adapt the agent plans to unexpected situations, or create a new plan to obtain the desired result in a different way.

These methods are reconfiguration methods and are therefore adapted to permanent failures, but they involve a higher cost because they imply some redundancy:

- To change its acquaintance, an agent needs to have other agents available capable of delivering the required service; this is agent redundancy.
- To change its plans, an agent needs to be able to contact some other available agents (know that they exist and how to join them) capable of delivering the required new services; this is service redundancy.

References

1. Arlat, J., Crouzet, Y., Deswarte, Y., Fabre, J.-C., Laprie, J.-C., Powell, D.: Tolérance aux fautes. In: Akoka, I.-W. J. (ed.) Encyclopédie de l'Informatique et des Systèmes d'Information, pp. 241–270. Vuibert, Paris (2006)
2. Avizienis, A.: Toward systematic design of fault-tolerant systems. Computer **30**(4), 51–58 (1997)

3. Avizienis, A., Laprie, J.-C., Randell, B., Landwehr, C.: Basic concepts and taxonomy of dependable and secure computing. IEEE Trans. Dependable Secure Comput. **1**(1), 11–33 (2004) (IEEE computer society, editor)
4. Barbuceanu, M., Fox, M.S.: Cool: a language for describing coordination in multiagent systems. In: Lesser, V., Gasser, L. (eds.) Proceedings of the first international conference oil multi-agent systems, pp. 17–24, AAAI Press, San Francisco (1995)
5. Castelfranchi, C., Falcone, R.: From automaticity to autonomy: the frontier of artificial agents. In: Hexmoore, H., Castelfranchi, C., Falcone, R. (ed.) Agent Autonomy, pp. 103–136. Kluwer Academic Publishers, Dordrecht (2003)
6. Chopinaud, C., Fallah-Seghrouchni, A.E., Taillibert, P.: Prevention of harmful behaviors within cognitive and autonomous agents. In: European conference on artificial intelligence, pp. 205–209 (2006)
7. Cristian, F., Fetzer, C.: The timed asynchronous distributed system model. In: FTCS, pp. 140–149 (1998)
8. Dastani, M.: 2APL: a practical agent programming language. Int. J. of Auton. Agent. Multi-Agent Syst. (JAAMAS). **16**(3), 214–248 Special Issue on Computational Logic-based Agents, Francesca Toni, Jamal Bentahar (eds.) (2008)
9. de Weerdt, M., ter Mors, A., Witteveen, C.: Multi-agent planning: An introduction to planning and coordination. In: Handouts of the European Agent Summer School, pp. 1–32 (2005)
10. d'Inverno, M., Luck, M.: Understanding autonomous interaction. In: Wahlster, W. (ed.) Proceedings of the 12th European Conference on Artificial Intelligence, pp. 529–533. John Wiley and Sons, Chichester (1996)
11. Dragoni, N., Gaspari, M.: Crash failure detection in asynchronous agent communication languages. Auton. Agent. Multi-Agent Syst. **13**(3), 355–390 (2006)
12. Fedoruk, A., Deters, R.: Improving fault-tolerance by replicating agents. In: Proceedings of the first international joint conference on autonomous agents and multiagent systems: part 2, pp. 737–744. ACM Press, Bologna, (2002)
13. Finin, T., Labrou, Y., Mayfield, J.: KQML as an agent communication language. In: Bradshaw, J. (ed.), Software Agents. MIT Press, Cambridge (1997)
14. D.T. FIPA.: FIPA communicative act library specification (2001)
15. Gartner, F.C.: Fundamentals of fault-tolerant distributed computing in asynchronous environments. ACM Comput. Surv. **32**(1), 1–26 (1999)
16. Guessoum, Z., Faci, N., Briot, J.-P.: Adaptive replication of large-scale multiagent systems: towards a fault-tolerant multi-agent platform. In: Proceedings of the fourth international workshop on software engineering for large-scale multi-agent systems, pp. 1–6. ACM Press, St. Louis (2005)
17. Hägg, S.: A sentinel approach to fault handling in multi-agent systems. In: Second Australian workshop on distributed AI in conjunction with the fourth Pacific rim international conference on artificial intelligence, pp. 181–195 (1996)
18. Hamming, R.W.: Error detecting and error correcting codes. Bell Syst. Tech. J. **26**(2), 147–160 (1950)
19. Hexmoor, H.: Stages of autonomy determination. IEEE Trans. Syst. Man Cybern. Part C **31**(4), 509–517 (2001) (IEEE computer society, editor)
20. Klein, M., Dellarocas, C.: Exception handling in agent systems. In: Etzioni, O., Müller, J.P., Bradshaw, J.M., (eds.) Proceedings of the third international conference on autonomous agents (Agents'99), pp. 62–68. ACM Press, Washington, (1999)
21. Koren, I., Koren, Z., Su, S.Y.: Analysis of a class of recovery procedures. IEEE Trans. Comput. **35**(8), 703–712 (1986)
22. Kuwabara, K.: Meta-level control of coordination protocols. In: Second international conference on multi-agent systems, pp. 165–172. (1996)
23. Laprie, J.-C., Béounes, C., Kanoun, K.: Definition and analysis of hardware and software-fault-tolerant architectures. Computer **23**(7), 39–51 (1990)

24. Nelson, V.P.: Fault-tolerant computing: fundamental concepts. Computer **23**(7), 19–25 (1990)
25. Platon, E., Sabouret, N., Honiden, S.: A definition of exceptions in agent oriented computing. In: O'Hare, G., O'Grady, M., Dikenelli, O., Ricci, A. (eds.) Engineering Societies in the Agent World'06 (2006)
26. Potiron, K., Taillibert, P., Fallah-Seghrouchni, A.E.: Autonomous agents: when the mailbox remains empty. In: IAT (2009)
27. Potiron, K., Taillibert, P., Fallah-Seghrouchni, A.E.: A new performative for handling lack of answers of autonomous agents. In: ICAART, pp. 441–446 (2009)
28. Potiron, K., Taillibert, P., Fallah-Seghrouchni, A.E.: A step towards fault tolerance for multi-agent systems. In: Languages, Methodologies and Development Tools for Multi-Agent Systems First International Workshop. Revised Selected Books Lecture Notes in Computer Science, vol. 5118, 4–6 Sept (2007)
29. Sabater, J., Sierra, C.: Review on computational trust and reputation models. Artif. Intell. Rev. **24**(1), 33–60 (2005)
30. Schneider, F.B.: Synchronization in distributed programs. ACM Trans. Program. Lang. Syst. **4**(2), 125–148 (1982)
31. Searle, J.R.: Speech Acts: An Essay in the Philosophy of Language. Cambridge University Press, Cambridge (1969)
32. El Fallah Seghrouchni, A., Hashmi M.A.: Multi-agent planning. In: Essaaidi, M. et al. (eds.) NATO Science for Peace and Security. Software Agents, Agent Systems and Their Applications. IOS Press, Amsterdam (2012)
33. Shah, N., Chao, K.-M., Godwin, N., James, A.E.: Exception diagnosis in open multi-agent systems. In IAT, pp. 483–486 (2005)
34. Wooldridge, M., Jennings, N.R.: Intelligent agents: theory and practice. Knowl. Eng. Rev. **10**, 115–152 (1995)
35. Zhang, Y., Manisterski, E., Kraus, S., Subrahmanian, V.S., Peleg, D.: Computing the fault tolerance of multi-agent deployment. Artif. Intell. **173**(3–4), 437–465 (2009)